A Formulary for Plasma Physics

by André Anders

Akademie-Verlag Berlin

TABLE OF PHYSICAL CONSTANTS

constant	*SI*	*CGS*
c	2.998×10^{8} m/s	2.998×10^{10} cm/s
e	1.602×10^{-19} As	4.803×10^{-10} statcoul
m_e	9.109×10^{-31} kg	9.109×10^{-28} g
m_u	1.661×10^{-27} kg	1.661×10^{-24} g
h	6.626×10^{-34} J s	6.626×10^{-27} erg s
k	1.381×10^{-23} J/K	1.381×10^{-16} erg/K
ε_0	8.854×10^{-12} As/Vm	
μ_0	$4\pi \times 10^{-7}$ Vs/Am	

A more detailed table of physical constants is to be found on pages 15 - 19.

André Anders

A Formulary for Plasma Physics

A Formulary for Plasma Physics

by André Anders

Akademie-Verlag Berlin

Author:
Dr. rer. nat. André Anders
Zentralinstitut für Elektronenphysik
Akademie der Wissenschaften
Berlin

ISBN 3-05-501263-1
Erschienen im Akademie-Verlag Berlin, Leipziger Str. 3–4, Berlin O-1086

Printed in Germany
Druck: Arthur Collignon GmbH, Berlin
Buchbinderische Verarbeitung: Peter u. Tomas Lehmann, Berlin
LSV 1197
Bestellnummer: 764 201 9 (9362)

PREFACE

The motivation to compile this formulary originated from the lack of a concise but comprehensive collection of useful expressions to which I could refer during my research in Plasma Physics. It should not be the case, after all, that a scientist should waste valuable time in a protracted search for the most appropriate formula. It is, therefore, the aim of this book to present a compilation of basic Plasma Physics formulae in such a way as to facilitate quick and easy location of the required expression. I have, in addition, attempted to include, in many cases, abbreviated or approximate forms which may be used for quick calculation, together with an extensive collection of references where the derivation and more detailed discussion of each formula may be found. Whilst this formulary cannot hope to encompass all theoretical work in this enormous and rapidly expanding field, I hope it will be a helpful tool to those of you who, like me, do not wish to spend more time searching for the equation than computing the answer.

After having completed half of this work, I received a booklet written by David L. Book and entitled *"NRL Plasma Formulary"* [35]. Although I discovered many of my own ideas already presented there, I was encouraged by my colleagues to continue and to complete my own version of a formulary. Having done so, I am, nevertheless, deeply indebted to Mr. Book who generously granted permission for unrestricted use of his material. In accordance with his wishes, I would like to draw the reader's attention to alternative formularies such as *"Astrophysical Quantities"* [78], by C.W. Allen, the *Physicist's Desk Reference"* [176], and *"Astrophysical Formulae"* by K.R. Lang [177].

It is also a pleasure to extend grateful acknowledgment to my colleagues Prof. Erhard Hantzsche, Prof. Burkhard Jüttner, Dr. Torsten Kahlbaum, Dr. Uwe Wenzel, Dr. Gottfried Hartel, Dr. Gerhard Babucke and to my wife, Dr. Simone Anders for their encouragement, support and helpful comments. Special thanks go to Dr. Richard A. Pitts, Culham Laboratory, U.K., and to Dr. Dirk Sünder for critical revision of the manuscript.

Finally, I would be very grateful to anyone who wishes to write concering errors in the text or with suggestions for improvement.

André Anders

Zentralinstitut für Elektronenphysik

Akademie der Wissenschaften

Hausvogteiplatz 5-7

Berlin 1086

Germany

HOW TO USE THIS BOOK

Frequently used symbols are compiled on pages 13 and 14. Additional, more specific notation is explained as it occurs in the text. The sign "→" refers to a formula or term occuring elsewhere in the book.
Although the "Système International d'Unités" , *SI*, is now established world-wide, the "Centimetre-Gramm-Second" System, *CGS*, also is still wideldy used. I have, therfore, tried to give the majority of formulae in both systems. Where an equation containing numerical factors occurs, a table of this kind

	phys. quantities
SI	units
CGS	units

gives the appropriate units for each physical quantity in the expression.

CONTENTS

FREQUENTLY USED SYMBOLS

a_0	Bohr radius
B	magnetic induction (magnetic flux density)
c	velocity of light in vacuum
c_p, c_V	specific heats at constant pressure and volume
C	capacitance
$\mathfrak{D}$	diffusion coefficient
D	eletric displacement
e	electronic charge
E	energy
E	electric field
f	frequency, distribution function
g	statistical weight, Gaunt factor
h	Planck constant ($\hbar = h / 2\pi$)
H	magnetic intensity
i	$= (-1)^{1/2}$
I	electric current
j	electric current density
k	Boltzmann constant, wave number

$$K_\varepsilon = \begin{cases} 1/(4\pi\, \varepsilon_0) & (SI) \\ 1 & (CGS) \end{cases}$$ arbitary constant for compact representation of formulae in both systems of units

l	length
L	inductance
m	mass
n	number density
N	number of particles
p	pressure
q	electric charge
r	radius
r_L	Larmor radius
Ry	Rydberg constant
t	time
T	temperature
u	velocity
v	velocity
V	potential, volume
x,y,z	cartesian coordinates
Z	ion charge number

Z_{max}	maximal ion charge number
Z_{eff}	effective nuclear charge
α	degree of ionization, recombination coefficient
α_F	fine structure constant
β	ionization coefficient
γ	parameter of a strongly coupled plasma
Γ	plasma parameter
ε_0	permittivity of vacuum
η	dynamic viscosity
$\varkappa$	thermal conductivity, ratio of specific heats
λ	wavelength
λ_D	Debye length
$\ln \Lambda$	Coulomb logarithm
μ	mobility, kinematic viscosity
μ_0	permeability of vacuum
ν	frequency
ξ	degeneration parameter
ρ	mass density, electric charge density
σ	cross-section, electrical conductivity, partition function
τ	characteristic time, period
ω	angular frequency
ω_c	cyclotron frequency
ω_{pl}	plasma frequeny

Indices

a	atom
b	bound
D	Debye
e	electron
f	field particle, free, flow
F	Fermi
h	heavy particle
H	hydrogen
i	ion
i, j	tensor indicies
l	lower level
pl	plasma
t	test particle
u	upper level
α	species of type α

1. PHYSICAL CONSTANTS

In this section, the 1986 set of recommended values of the basic constants and conversion factors is given, published by CODATA (e.g.[1]), the committee on Data for Science and Technology of the International Council of Scientific Unions. Useful relations between fundamental and derived constants are also included.

physical quantity	symbol, relations	mantissa	order of magnitude, unit: *CGS*	order of magnitude, unit: *SI*	relative uncert. in ppm
speed of light in vacuum	c	2.99792458	10^{10}cm/s	10^{8}m/s	exact
permeability of vacuum	μ_0	4π		10^{-7} Vs/Am	exact
permittivity of vacuum	$\varepsilon_0 = 1 / \mu_0 c^2$	8.854187817		10^{-12} As/Vm	exact
elementary charge	e	1.60217733		10^{-19}As	0.30
		4.80324	10^{-10}stat-coulomb		
electron mass	m_e	9.1093897	10^{-28}g	10^{-31}kg	0.59
proton mass	m_p	1.6726231	10^{-24}g	10^{-27}kg	0.59
neutron mass	m_n	1.6749286	10^{-24}g	10^{-27}kg	0.59
deuteron mass	m_d	3.3435860	10^{-24}g	10^{-27}kg	0.59

physical quantity	symbol, relations	mantissa	order of magnitude, unit CGS	SI	relative uncert. in ppm
atomic mass unit	m_u	1.6605402	10^{-24}g	10^{-27}kg	0.59
Planck constant	h	6.6260755	10^{-27}erg s	10^{-34}J s	0.60
	$\hbar=h/2\pi$	1.05457266			
Boltzmann constant	k_B	1.380658	10^{-16}erg/K	10^{-23}J/K	8.5
Avogadro constant	N_A	6.0221367	10^{23}mol^{-1}	10^{23} mol^{-1}	0.59
Loschmidt constant	$L_0 = N_A/V_m$	2.686763	10^{19}cm^{-3}	10^{25} m^{-3}	8.5
universal molar gas constant	$R = N_A\ k_B$	8.314510	10^7 erg/K mol	J/K mol	8.4
standard temperature	T_0	273.15	K	K	exact
standard pressure	p_0 = 1 atm	1.01325	10^6 dyne/cm^2	10^5Pa	exact
molar volume (ideal gas)	$V_m = R\ T_0/p_0$	2.241410	10^4cm^3/mol	10^{10} m^3/mol	
Faraday constant	$F = N_A\ e$	9.6485309		10^4 $\frac{As}{mol}$	0.30
		2.89253	10^{14} stat-coul./mol		

physical quantity	symbol, relations	mantissa	order of magnitude, unit *CGS*	*SI*	relative uncert. in ppm
fine-structure constant	$\alpha_F = e^2 K_\varepsilon / \hbar c$	7.29735308	10^{-3}	10^{-3}	0.045
	α_F^{-1}	137.0359895			
Rydberg constant in wavenumber units	$Ry_\infty = K_\varepsilon^2 2\pi^2 m_e e^4 / ch^3$	1.0973731534	$10^5\ cm^{-1}$	$10^7\ m^{-1}$	0.0012
in energy units	$K_\varepsilon^2 2\pi^2 m_e e^4 / ch^2$	2.1798741	10^{-11} erg	10^{-18} J	0.60
		13.6056981	eV	eV	
in frequency units	$K_\varepsilon^2 2\pi^2 m_e e^4 / h^3$	3.2898419499	$10^{15}\ s^{-1}$	$10^{15}\ s^{-1}$	0.0012
Bohr radius	$a_0 = \hbar^2 / K_\varepsilon e^2 m_e = \alpha / 4\pi\ Ry_\infty$	5.29177249	10^{-9} cm	10^{-11} m	0.045
atomic cross-section	$\pi\ a_0^2$	8.797356696	$10^{-17}\ cm^2$	$10^{-21}\ m^2$	
Compton wavelength of electrons	$\lambda_C = h/m_e\ c = \alpha^2/2\ Ry_\infty$	2.42631058	10^{-10}cm	10^{-12}m	0.089
	$\bar{\lambda}_C = \lambda_C / 2\pi = \alpha\ a_0$	3.86159323	10^{-11}cm	10^{-13}m	

physical quantity	symbol, relations	mantissa	order of magnitude, unit CGS	SI	relative uncert. in ppm
classical electron radius	$r_e = \alpha \lambda_C = \alpha^2 a_0 = e^2 K_\varepsilon / c^2 m_e$	2.81794092	10^{-13} cm	10^{-15} m	0.13
Thomson cross section	$\sigma_e = (8\pi/3)\ r_e^2$	6.66524616	10^{-25} cm^2	10^{-29} m^2	0.27
Bohr magneton	$\mu_B =$ *SI:* $e\hbar/2m_e$ *CGS:* $e\hbar/2cm_e$	9.2740154	10^{-21} erg/gauss	10^{-24}J/T	0.34
nuclear magneton	$\mu_N =$ *SI:* $e\hbar/2m_p$ *CGS:* $e\hbar/2cm_p$	5.0507866	10^{-24} erg/gauss	10^{-27}J/T	0.34
magnetic moment of the electron	$\mu_e = 1.00116\ \mu_B$	9.2847701	10^{-21} erg/gauss	10^{-24}J/T	0.34
magnetic moment of the proton	μ_p	1.41060761	10^{-23} erg/gauss	10^{-26}J/T	0.34
magnetic moment of the neutron	μ_n	9.6623707	10^{-24} erg/gauss	10^{-27}J/T	0.34

physical quantity	symbol, relations	mantissa	order of magnitude, unit CGS	SI	relative uncert. in ppm
magnetic moment of the deuteron	μ_d	4.3307375	10^{-24} erg/gauss	10^{-27} J/T	0.34
Stefan-Boltzmann constant	$\sigma_{St/B} = 2\pi^5 k_B^4 / 15\ c^2 h^3$	5.67051	10^5 $\frac{erg}{s\ cm^2\ K^4}$	10^{-8} W/m^2K^4	34
first radiation constant	$c_1 = 2\pi\ h\ c^2$	3.7417749	10^{-5} erg cm^2/s	10^{-8} W m^2	0.60
second radiation constant	$c_2 = h\ c\ /\ k$	1.438769	cm K	10^{-2} m K	8.4

2. PHYSICAL QUANTITIES, THEIR DIMENSIONS AND UNITS

Although the "Systeme International d'Unites", *SI*, is now introduced world-wide, the "Centimetre-Gramm-Second" System, *CGS*, is still widely used. In *CGS* systems, both mechanical and electromagnetic quantities can be expressed in the form

$$cm^p \, g^q \, s^r$$

where p, q, r are positive or negative, integer or rational numbers. In the equations of the

* energy density of the electric field $\qquad \mathscr{E}_E = \frac{1}{2} \varepsilon E^2 \qquad (1)$

* energy density of the magnetic field $\qquad \mathscr{E}_H = \frac{1}{2} \mu H^2 \qquad (2)$

* energy flux density $\qquad S = \gamma \, E \times H \qquad (3)$

the interelated factors ε, μ, and γ occur. Generally these factors depend on the substance under consideration; the subscript "0" refers to a value in vacuum. Two of the factors can be choosen arbitrarily. Different *CGS*-systems can be distinguished:

(1) Gaussian system

Here, $4\pi \, \varepsilon_0^{Gauss} = 1$ and $\dfrac{\mu_0^{Gauss}}{4\,\pi} = 1$.

Then, $\left[E^{Gauss}\right] = \left[H^{Gauss}\right] = (\text{energy density})^{1/2} = \left(\dfrac{cm^2 \, g \, s^{-2}}{cm^3}\right)^{1/2}$

$$= cm^{-1/2} \, g^{1/2} \, s^{-1}$$

and, with equation (3), $\left[\gamma^{Gauss}\right] = cm \, s^{-1}$ i.e. $\gamma^{Gauss} = c$, the velocity of light in vacuum.

(2) electrostatic cgs (esu)

Here, $4\pi \, \varepsilon_0^{esu} = 1$ and $\gamma^{esu} = 1$.

Then, $\left[E^{esu}\right] = cm^{-1/2} \, g^{1/2} \, s^{-1}$, $\left[H^{esu}\right] = cm^{1/2} \, g^{1/2} \, s^{-2}$,

and $\left[\mu_0^{esu}\right] = s^2 \, cm^{-2}$ i.e. $\dfrac{\mu_0^{esu}}{4\,\pi} = 1/c^2$.

(3) electromagnetic cgs (emu)

Here, $\frac{\mu_0^{emu}}{4\pi} = 1$ and $\gamma^{emu} = 1$.

Then, $\left[E^{emu}\right] = cm^{1/2}\, g^{1/2}\, s^{-2}$, $\left[H^{emu}\right] = cm^{-1/2}\, g^{1/2}\, s^{-1}$,

and $\left[4\pi\, \varepsilon_0^{emu}\right] = s^2\, cm^{-2}$ i.e. $4\pi\, \varepsilon_0^{emu} = 1/c^2$.

The relations between electrical and magnetic quantities in different systems of units are given in the table at the next page. For precise work, the factors 3 and 9 should be replaced by 2.9979 and 8.9875, respectively.

In *CGS*, the Coulomb law $F = K_\varepsilon\, q_1 q_2/r^2$ is used to define the unit of electric charge (1 el.stat. charge unit), when $K_\varepsilon = 1$, $r = 1$ cm, $F = 1$ dyn. Since the unit of electric charge in *SI* (1 Coulomb) is defined in terms of the force between current carring conductors, we have the relation

$$1 \text{ Coulomb} = 2.9979 \times 10^9 \text{ el. stat. charge units}$$

The numerical factor is one tenth of the velocity of light in vacuum. For the constant of proportionality this definition gives

$$K_\varepsilon = \frac{(1\text{ dyn})\ \ (1\text{ cm})^2}{(1\text{ el.stat.ch.u.})^2} = \frac{10^{-5}\text{N}\ \ (10^{-2}\text{m})^2}{(1\text{ C} / 2.998 \times 10^9)^2}$$

$$= 8.988 \times 10^9 \text{ N m}^2/\text{C}^2 \cong 9 \times 10^9 \text{ N m}^2/\text{C}^2 = 9 \times 10^9 \text{ Vm/As}$$

$$= 1/(4\pi\, \varepsilon_0)$$

In order to use the formulae given in this book in both systems of units, we will often use

$$K_\varepsilon = \begin{cases} 1\ /(4\pi\, \varepsilon_0) = 8.988 \times 10^9 \text{ Vm/As} & (SI) \\ 1 & (CGS) \end{cases}$$

quantity	*SI*	Gaussian	esu	emu
voltage U	1 Volt	$\frac{1}{300}\ \mathrm{cm}^{\frac{1}{2}}\ \mathrm{g}^{\frac{1}{2}}\ \mathrm{s}^{-1}$	$\frac{1}{300}\ \mathrm{cm}^{\frac{1}{2}}\ \mathrm{g}^{\frac{1}{2}}\ \mathrm{s}^{-1}$	$10^{8}\ \mathrm{cm}^{\frac{3}{2}}\ \mathrm{g}^{\frac{1}{2}}\ \mathrm{s}^{-2}$
current I	1 Ampere	$3\times10^{9}\ \mathrm{cm}^{\frac{3}{2}}\ \mathrm{g}^{\frac{1}{2}}\ \mathrm{s}^{-2}$	$3\times10^{9}\ \mathrm{cm}^{\frac{3}{2}}\ \mathrm{g}^{\frac{1}{2}}\ \mathrm{s}^{-2}$	$0.1\ \mathrm{cm}^{\frac{1}{2}}\ \mathrm{g}^{\frac{1}{2}}\ \mathrm{s}^{-1}$
charge q	1 Coulomb	$3\times10^{9}\ \mathrm{cm}^{\frac{3}{2}}\ \mathrm{g}^{\frac{1}{2}}\ \mathrm{s}^{-1}$	$3\times10^{9}\ \mathrm{cm}^{\frac{3}{2}}\ \mathrm{g}^{\frac{1}{2}}\ \mathrm{s}^{-1}$	$0.1\ \mathrm{cm}^{\frac{1}{2}}\ \mathrm{g}^{\frac{1}{2}}$
resistance R	1 Ohm	$\frac{1}{9\times10^{11}}\ \mathrm{cm}^{-1}\ \mathrm{s}$	$\frac{1}{9\times10^{11}}\ \mathrm{cm}^{-1}\ \mathrm{s}$	$10^{9}\ \mathrm{cm\ s}^{-1}$
capacitance C	1 Farad	$9\times10^{11}\ \mathrm{cm}$	$9\times10^{11}\ \mathrm{cm}$	$10^{-9}\ \mathrm{cm}^{-1}\ \mathrm{s}^{2}$
inductance L	1 Henry	$\frac{1}{9\times10^{11}}\ \mathrm{cm}^{-1}\ \mathrm{s}^{2}$	$\frac{1}{9\times10^{11}}\ \mathrm{cm}^{-1}\ \mathrm{s}^{2}$	$10^{9}\ \mathrm{cm}$
electric field E	1 V/m	$\frac{1}{3\times10^{4}}\ \mathrm{cm}^{-\frac{1}{2}}\ \mathrm{g}^{\frac{1}{2}}\mathrm{s}^{-1}$	$\frac{1}{3\times10^{4}}\ \mathrm{cm}^{-\frac{1}{2}}\ \mathrm{g}^{\frac{1}{2}}\mathrm{s}^{-1}$	$10^{6}\ \mathrm{cm}^{\frac{1}{2}}\ \mathrm{g}^{\frac{1}{2}}\ \mathrm{s}^{-2}$
electric dis- D placement	1 $\mathrm{C/m}^2$	$4\pi\times3\times10^{5}$ $\mathrm{cm}^{-1/2}\ \mathrm{g}^{1/2}\ \mathrm{s}^{-1}$	$4\pi\times3\times10^{5}$ $\mathrm{cm}^{-1/2}\ \mathrm{g}^{1/2}\ \mathrm{s}^{-1}$	$4\pi\times10^{-5}\mathrm{cm}^{-\frac{3}{2}}\ \mathrm{g}^{\frac{1}{2}}$
magnetic field H	1 $\mathrm{A/m}^2$	$4\pi\times3\times10^{7}$ $\mathrm{cm}^{-1/2}\ \mathrm{g}^{1/2}\ \mathrm{s}^{-1}$	$4\pi\times3\times10^{7}$ $\mathrm{cm}^{1/2}\ \mathrm{g}^{1/2}\ \mathrm{s}^{-2}$	$\frac{4\pi}{10^{3}}\ \mathrm{cm}^{-\frac{1}{2}}\ \mathrm{g}^{\frac{1}{2}}\ \mathrm{s}^{-1}$
magnetic induc- B tance	1 $\mathrm{Weber/m}^2$	$10^{4}\ \mathrm{cm}^{-\frac{1}{2}}\ \mathrm{g}^{\frac{1}{2}}\ \mathrm{s}^{-1}$	$\frac{1}{3\times10^{4}}\ \mathrm{cm}^{\frac{3}{2}}\ \mathrm{g}^{\frac{1}{2}}$	$10^{4}\ \mathrm{cm}^{-\frac{1}{2}}\ \mathrm{g}^{\frac{1}{2}}\ \mathrm{s}^{-1}$

Unless specified otherwise, we will use the Gaussian cgs, denoted "*CGS*".

Reference: [122], appendix.

Physical quantities, their dimension and units

physical quantity	symbol	dimension SI	dimension CGS	units SI	units CGS	conversion
		mechanics and thermodynamics				
length	l	l	l	m meter	cm centimeter	1 m = 10^2 cm = 10^{10} Å
area	A	l^2	l^2	m^2	cm^2	1 m^2 = 10^4 cm^2 = 10^{28} barn
volume	V	l^3	l^3	m^3	cm^3	1 m^3 = 10^6 cm^3
number density	n	l^{-3}	l^{-3}	m^{-3}	cm^{-3}	1 m^{-3} = 10^{-6} cm^{-3}
plane angle	θ	l/l	l/l	rad	rad	$1^\circ = \pi$ / 180 rad 1′ = 1°/60 =2.90888×10^{-4}rad 1′′=1′/60 =4.8481×10^{-6}rad
solid angle	Ω	l^2/l^2	l^2/l^2	sr	sr	1 sr = = 1m/1m = 1cm/1cm
time	t	t	t	s second	s second	1 s = 1/60 min = 1/3600 h 1 d = 86400 s 1 a = 365.2422 d = 3.1557×10^7s
frequency	f ν	t^{-1}	t^{-1}	s^{-1}	s^{-1}	1 s^{-1} = = 1 Hz (Hertz)
velocity	v	l/t	l/t	m/s	cm/s	1 m/s = 100 cm/s = 3.6 km/h

physical quantity	symbol	dimension SI	dimension CGS	units SI	units CGS	conversion
acceleration	a	$1/t^2$	$1/t^2$	m/s^2	cm/s^2	$1\ m/s^2 = 100\ cm/s^2$
mass	m	m	m	kg kilogram	g gram	1 kg = 1000 g
mass density	ρ	m/l^3	m/l^3	kg/m^3	g/cm^3	$1\ kg/m^3 = 10^{-3} g/cm^3$
force	F	ml/t^2	ml/t^2	N Newton	dyn dyne	$1\ N = 1\ kg\ m/s^2$ $= 10^5$ dyn $1\ dyn = 1\ g\ cm/s^2$ 1 kp = 9.80665 N
pressure, tension	p	m/lt^2	m/lt^2	Pa Pascal	dyn/cm^2	1 Pa = $= 1 N/m^2 = 1 kg/m\ s^2$ $= 10\ dyn/cm^2$ $1\ dyn/cm^2$ $= 1\ g/cm\ s^2$
						1 Torr = 1 mm Hg = 133.3224 Pa $1\ kp/m^2$ = 9.80665 Pa 1 at = $1\ kp/cm^2 = 9.80665 \times 10^4$ Pa 1 atm = 760 Torr = 1.01325×10^5 Pa 1 bar = 10^5 Pa
energy, enthalpy, work, heat	E H W Q	ml^2/t^2	ml^2/t^2	J Joule	erg	1 J = 1 Nm = 1 Ws $= 1 VAs = 1\ kg\ m^2/s^2$ $= 10^7$ erg $1\ erg = 1\ g\ cm^2/s^2$ 1 cal = 4.1868 J 1 eV = $1.60217733 \times 10^{-19}$ J $1.6021773 \times 10^{-12}$ erg $\hat{=}$ 11605 K $\hat{=}$ 8065.541 cm^{-1}

physical quantity	symbol	dimension SI	dimension CGS	units SI	units CGS	conversion
power	P	ml^2/t^3	ml^2/t^3	W Watt	erg/s	1 W = 1 J/s = 1 VA $= 1$ kg m^2/s^3 $= 10^7$erg/s
momentum	p	ml/t	ml/t	kg m/s	g cm/s	1 kg m/s = $= 10^5$g cm/s
angular momentum	l	ml^2/t	ml^2/t	kg m^2/s	g cm^2/s	1 kg m^2/s = 1 Js $= 10^7$g cm^2/s
dynamic viscosity	η	m/lt	m/lt	Pa s	P Poise	1 Pa s = 1 N s/m^2 = 1 kg/s m = 10 P 1 P = 1 g/s cm
kinematic viscosity	μ	l^2/t	l^2/t	m^2/s	St Stokes	1 m^2/s = 10^4 St $= 10^4$ cm^2/s
temperature	T	T	T	K Kelvin	K Kelvin	T [°C] = T [K]-273.15 K
entropy	S	ml^2/t^2T	ml^2/t^2T	J/K	erg/K	1 J/K = 10^7 erg/K
thermal conductivity	$\varkappa$	ml/t^3T	ml/t^3T	W/m K	erg/s cm K	1 W/m K = = 1 kgm/s^3K $= 10^5$erg/s cm K 1 erg/s cm K = = 1 g cm/s^3 K
diffusion coeff.	$\mathfrak{D}$	l^2/t	l^2/t	m^2/s	cm^2/s	1 m^2/s = 10^4 cm^2/s
specific heat capacity	c	l^2/t^2T	l^2/t^2T	J/kg K	erg/g K	1J/kg K = 1 m^2/s^2K $= 10^4$ erg/g K $= 10^4cm^2/s^2K$

physical quantity	symbol	dimension SI	dimension CGS	units SI	units CGS	conversion
		electricity and magnetism				
charge	q	q	$\frac{m^{1/2} l^{3/2}}{t}$	C Coulomb	stat-coulomb	1 C = 1 As =2.998×10^9statcoul 1 As =6.2415×10^{18}e 1 statcoul. = = 2.0805×10^9e
electric current intensity	I	q/t	$\frac{m^{1/2} l^{3/2}}{t^2}$	A Ampere	stat-ampere	1 A = =2.998×10^9 statamp
current density	j	q/l^2t	$\frac{m^{1/2}}{l^{1/2}\,t^2}$	A/m^2	$\frac{\text{statampere}}{cm^2}$	$1\ A/m^2 = 10^{-4} A/cm^2$ $= 3\times10^5\ \frac{\text{statampere}}{cm^2}$
electric potential voltage	Φ V U	$\frac{m\,l^2}{t^2\,q}$	$\frac{m^{1/2} l^{1/2}}{t}$	V Volt	statvolt	$1\ V = 1\ kg\ m^2/A\ s^3$ $= \frac{10^{-2}}{2.998}$ statvolt 1 statvolt = 1 g cm^2/s^2statcoul
electric displacement, polarization	D P	q/l^2	$\frac{m^{1/2}}{l^{1/2}t}$	C/m^2	$\frac{\text{statcoulomb}}{cm^2}$	$1\ C/m^2 = 1\ As/m^2 =$ $= 12\pi\ 10^5\ \frac{\text{statcoul}}{cm^2}$ $= 10^{-4}\ As/cm^2$
electric field	E	$\frac{m\,l}{t^2\,q}$	$\frac{m^{1/2}}{l^{1/2}\,t}$	V/m	statvolt/cm	$1\ V/m = 1\ kg\ m/s^3A$ $= \frac{10^{-4}}{2.998}\ \frac{\text{statvolt}}{cm}$ $= 10^{-2}$ V/cm

physical quantity	symbol	dimension SI	dimension CGS	units SI	units CGS	conversion
reduced electric field	$\frac{E}{n}$	$\frac{l^4}{t^2\,q}$	$\frac{m^{1/2}l^{5/2}}{t}$	Td Townsend	Td Townsend	1 Td = 10^{-21} V m^2 = 10^{-17} V cm^2 =10^{-15}statvolt cm^2
capacitance	C	$\frac{t^2q^2}{m\,l^2}$	l	F Farad	cm	1 F = 1 As/V = 1 A^2s^4/kg m^3 $(2.998)^2\times 10^{11}$cm
resistance / impedance	R / Z	$\frac{m\,l^2}{t\,q^2}$	t/l	Ω Ohm	s/cm	1 Ω = 1 V/A = 1 kg m^2/s^3A^2 $\frac{10^{-11}}{(2.998)^2}$ s/cm
electrical conductivity	σ	$\frac{t\,q}{m\,l^2}$	t^{-1}	$(\Omega\ m)^{-1}$	s^{-1}	1 $(\Omega\ m)^{-1}$= 1 A/Vm = 1 S (Siemens)/m = 1 A^2s^3/kg m^2 = 1 mho / m $(2.998)^2\times 10^9$ s^{-1}
magnetic intensity / magnetization	H / M	q/lt	$\frac{m^{1/2}}{l^{1/2}\,t}$	A/m	Oe Oersted	1 A/m = = $4\pi\times 10^{-3}$ Oe
magnetic flux	Φ	ml^2/tq	$\frac{m^{1/2}l^{3/2}}{t\,q}$	Wb Weber	M Maxwell	1 Wb= 1 Vs = 10^8 M = 1 kg m^2/s^2 A
magnetic induction (magnetic flux density)	B	m/tq	$\frac{m^{1/2}}{l^{1/2}t}$	T Tesla	G Gauss	1 T = 1 Wb/m^2 = $1Vs/m^2$ = $1kg/s^2A$ =10^4 G =10^4 M/cm^2

physical quantity	symbol	dimension SI	dimension CGS	units SI	units CGS	conversion
inductance	L	ml^2/q^2	t^2/l	H Henry	s^2/cm	1 H = 1 Wb/A $= 1kg\ m/s^2A^2$ $\frac{10^{-11}}{(2.998)^2}\ s^2/cm$
permeability	μ	$m\ l/q^2$	1	H/m	1	1 H/m = 1 Wb/Am $= 1\ kg\ m/s^2\ A^2$ $=(1/4\pi)\ 10^7$ (CGS)
permittivity	ε	$\frac{t^2\ q^2}{m\ l^3}$	1	F/m	1	1 F/m = 1 As/Vm $= 1\ s^4A^2/kg\ m^3$
vector potential	A	ml/tq	$\frac{m^{1/2}l^{1/2}}{t}$	Wb/m	M/cm	1 Wb/m $\hat{=}$ 1 Vs/m $= 1\ kg\ m/s^2A$ = 1 T m $=10^6 M/cm = 10^6 G\ cm$
mobility	μ	qt/m	$\frac{l^{3/2}}{m^{1/2}}$	m^2/Vs	$\frac{cm^2}{statvolt\ s}$	$1m^2/Vs = 10^4cm^2/Vs$ $= 2.998 \times 10^6\ \frac{cm^2}{statvolt\ s}$
			photometric quantities			
luminous intensity	I	I	I	cd candela	cd candela	1cd= 0.9814 stand. candle (USA) = 0.94242 candle (USSR) = 1.107 Hefner candle
luminance	L	I/l^2	I/l^2	cd/m^2	sb stilb	$1\ cd/m^2=10^{-4}cd/cm^2$ $= 10^{-4}$ sb $= \pi\ 10^{-4}$ lambert $= \pi$ apostilb

physical quantity	symbol	dimension SI	dimension CGS	units SI	units CGS	conversion
luminous flux	Φ	$I\ l^2/l^2$	$I\ l^2/l^2$	lm lumen	cd sr	1 lm = 1 cd sr
quantity of light	Q	Itl^2/l^2	Itl^2/l^2	lm s	cd sr s	1 lm s = 1 cd sr s
illu-minance	E	$I\ l^2/l^4$	$I\ l^2/l^4$	lx lux	ph phot	1 lx = 1 lm/m^2 = 1 cd sr/m^2 = 10^{-4} lm/cm^2 = 10^{-4} ph = 10^{-4} sb sr = 1.11 × 10^{-4} Hefner phot = 0.09294 foot candle (USA)
exposure (quantity of illu-minance)	H	Itl^2/l^4	Itl^2/l^4	lx s	ph s	1 lx s = 1 lm s/m^2 = 1 cd sr s/m^2 = 10^{-4} ph s = 10^{-4} sb sr s = 1.11 × 10^{-4} Hefner phot second
		energetic quantities of radiation				
radiant flux (radiant power)	Φ	ml^2/t^3	ml^2/t^3	W	erg/s	1 W = 1 J/s = 10^7 erg/s

physical quantity	symbol	dimension SI	dimension CGS	units SI	units CGS	conversion
spectral (mono- chrom.) radiant flux	Φ_λ	ml/t^3	ml/t^3	W/nm	erg/s Å	1 W/m = 1 kg m/s^3 = 10^{-9} W/nm = 10^5 erg/s cm = 10^{-3} erg/s Å
	Φ_ν	ml^2/t^2	ml^2/t^2	W/Hz	erg/s Hz	1 W/Hz = 1 J = 10^7erg = $10^7 \frac{erg}{s\ Hz}$
radiant flux density, Pointing vector	φ S	m/t^3	m/t^3	W/m^2	erg/s cm^2	1 W/m^2 = 1 J/s m^2 = 1 kg/s^3 = 10^3 erg/s cm^2
radiant exposure, irradi- ation	H	m/t^2	m/t^2	J/m^2	erg/cm^2	1 J/m^2 = 1 kg/s^2 = 10^3 erg/cm^2
inte- grated radiation density	u	m/lt^2	m/lt^2	J/m^3	erg/cm^3	1 J/m^3 = = 10 erg/cm^3
spectral (mono- chrom.) radiation density	u_λ	m/l^2t^2	m/l^2t^2	J/m^3 nm	erg/cm^3 Å	1 J/m^4 = = 10^{-9} J/m^3 nm = 10^{-9} erg/cm^3 Å
	u_ν	m/lt	m/lt	J/m^3 Hz	erg/cm^3 Hz	1 J/m^3 Hz = = 10 erg/cm^3 Hz
intensity of radiation	I	m/lt^3	m/lt^3	W/m^3	erg/cm^3 s	1 W/m^3 = 1 J/s m^3 = 1 kg/m s^3 = 10 erg/cm^3 s

physical quantity	symbol	dimension SI	dimension CGS	units SI	units CGS	conversion
radiant intensity	I	$\frac{m\ l^4}{t^3\ l^2}$	$\frac{m\ l^4}{t^3\ l^2}$	W/sr	erg/s sr	1 W/sr = 1 J/s sr $= 1\ kg\ m^2/s^3\ sr$ $= 10^7$ erg/s sr
spectral (mono-chrom.) radiant intensity	I_λ	$\frac{m\ l^3}{t^3 l^2}$	$\frac{m\ l^3}{t^3 l^2}$	W/sr nm	erg/s sr Å	1 W/sr m = $= 1\ kg\ m/s^3\ sr$ $=10^{-9}$ W/sr nm $=10^5$ erg/s sr cm $=10^{-3}$ erg/s sr Å
	I_ν	$\frac{m\ l^4}{t^2\ l^2}$	$\frac{m\ l^4}{t^2\ l^2}$	W/sr Hz	erg/s sr Hz	1 W/sr Hz = 1 J/sr $= 10^7$ erg/sr $= 10^7$ erg/s sr Hz
radiant intensity per unit area also: radiance luminosity	L B	$\frac{m\ l^2}{t^3\ l^2}$	$\frac{m\ l^2}{t^3\ l^2}$	$W/sr\ m^2$	$\frac{erg}{s\ sr\ cm^2}$	$1\ W/sr\ m^2 =$ $= 1\ kg/s^3\ sr$ $= 1\ J/s\ sr\ m^2$ $= 10^3\ erg/s\ sr\ cm^2$
spectral radiant intensity per unit area	L_λ B_λ	$\frac{m\ l^2}{t^3\ l^3}$	$\frac{m\ l^2}{t^3\ l^3}$	$\frac{W}{sr\ m^2\ nm}$	$\frac{erg}{s\ sr\ cm^2\ \AA}$	$1\ W/sr\ m^2\ m =$ $= 10^{-9}\ W/sr\ m^2\ \AA$ $=10\ erg/s\ sr\ cm^2 cm$ $=10^{-7} erg/s\ sr\ cm^2 \AA$
	L_ν B_ν	$\frac{m\ l^2}{t^2\ l^2}$	$\frac{m\ l^2}{t^2\ l^2}$	$\frac{W}{sr\ m^2\ Hz}$	$\frac{erg}{s\ sr\ cm^2\ Hz}$	$1\ W/sr\ m^2 Hz =$ $= 1\ J/sr\ m^2$ $=10^3\ erg/sr\ cm^2$ $=10^3 erg/s\ sr\ cm\ Hz$

physical quantity	symbol	dimension SI	dimension CGS	units SI	units CGS	conversion
emission coefficient	ε	$\frac{m\ l^2}{t^3\ l^3}$	$\frac{m\ l^2}{t^3\ l^3}$	$W/sr\ m^3$	$\frac{erg}{s\ sr\ cm^3}$	$1\ W/sr\ m^3 =$ $= 1\ J/s\ sr\ m^3$ $= 10\ erg/s\ sr\ cm^3$
spectral emission coefficient	ε_λ	$\frac{m\ l^2}{t^3\ l^4}$	$\frac{m\ l^2}{t^3\ l^4}$	$\frac{W}{sr\ m^3\ nm}$	$\frac{erg}{s\ sr\ cm^3\ Å}$	$1\ W/sr\ m^3\ nm$ $= 10^9\ W/sr\ m^4$ $= 1\ erg/s\ sr\ cm^3\ Å$ $= 10^8\ erg/s\ sr\ cm^4$
	ε_ν	$\frac{m\ l^2}{t^2\ l^3}$	$\frac{m\ l^2}{t^2\ l^3}$	$\frac{W}{sr\ m^3\ Hz}$	$\frac{erg}{s\ sr\ cm^3\ Hz}$	$1\ W/sr\ m^3 Hz =$ $= 1\ J/sr\ m^3$ $= 1\ kg/m\ s^2\ sr$ $=10\ erg/s\ sr\ cm^3 Hz$
absorption coeff.	$\varkappa$	l^{-1}	l^{-1}	m^{-1}	cm^{-1}	$1\ m^{-1} = 10^{-2}\ cm^{-1}$
		radioactivity				
activity	A	t^{-1}	t^{-1}	Bq Becquerel	Ci Curie	1 Bq = 1 count/s 1 Ci = $= 3.7 \times 10^{10}$ Ci
absorbed radiation dose	D	l^2/t^2	l^2/t^2	rad (also: rd, rem)	Gy Gray	1 Gy = 1 J/kg 1 rad = 10^{-2} Gy = $= 10^2$ erg/g

physical quantity	symbol	dimension *SI*	dimension *CGS*	units *SI*	units *CGS*	conversion
		molar quantities				
amount of substance	n	n	n	mol	mol	1 mol = $= 6.022045 \times 10^{23}$ particles
molar mass	M	m/n	m/n	kg/mol	g/mol	1 kg/mol = $= 10^3$ g/mol
molar volume	V	l /n	l /n	m /mol	cm /mol	1 m^3/mol $= 10^6$ cm^3/mol
molar heat capacity	c	$\frac{m\ l^2}{t^2\ n\ T}$	$\frac{m\ l^2}{t^2\ n\ T}$	J/mol K	erg/mol K	1 J/mol K = $= 1$ kg m^2/s^2mol K $= 10^7$ erg/mol K

Metric Prefixes

multiple	prefix	symbol
10^{-18}	atto	a
10^{-15}	femto	f
10^{-12}	pico	p
10^{-9}	nano	n
10^{-6}	micro	μ
10^{-3}	milli	m
10^{3}	kilo	k
10^{6}	mega	M
10^{9}	giga	G
10^{12}	tera	T
10^{15}	peta	P
10^{18}	exa	E

3. FUNDAMENTAL DEFINITIONS AND RELATIONS

3.1. LENGTHS

3.1.1. DEBYE LENGTH (DEBYE RADIUS. SCREENING LENGTH)

Under the assumptions

(i) charged particles of each type α are Boltzmann distributed with a temperature T_α,

(ii) $|Z_\alpha e V| \ll kT_\alpha$, where V represents the electrostatic potential and Z_α is the charge of species α,

(iii) microscopic variations of potential resulting from the discrete nature of the particles may be neglected,

the Debye length is defined by

$$\lambda_D = \left[4\pi K_\varepsilon e^2 \sum_\alpha \frac{Z_\alpha^2 n_\alpha}{k T_\alpha} \right]^{-1/2}$$

where α = e, i and $Z_i = 0,1,\ldots,Z_{max}$.

For isothermal plasmas, $T_\alpha = T \;\; \forall\, \alpha$, the Debye length may be rewritten, as follows,

$$\lambda_D = \left[\frac{k T}{4\pi K_\varepsilon e^2 n_e (1 + \langle Z^2\rangle/\langle Z\rangle)} \right]^{1/2}$$

with the definitions

$$\langle Z^2\rangle = \frac{\sum Z^2 n_i^{z+}}{\sum n_i^{z+}} \quad \text{and} \quad \langle Z\rangle = \frac{\sum Z\, n_i^{z+}}{\sum n_i^{z+}}$$

and using the condition of quasi-neutrality $n_e = \sum_{Z=1}^{Z_{max}} Z\, n_i^{z+} = \langle Z\rangle\, n_i$.

Often, $\langle Z^2\rangle/\langle Z\rangle$ is simply replaced by Z and denoted as the "mean ion charge". This is, strictly speaking, correct only for approximate formulae.

If the plasma contains only singly charged ions,

$$\lambda_D = \left[\frac{k T}{8\pi K_\varepsilon e^2 n_e} \right]^{1/2}$$

Shielding length for ions by electrons (electronic Debye length):

$$\lambda_{D,e} = \left[\frac{k\,T_e}{4\pi\,K_\varepsilon\,e^2\,n_e} \right]^{1/2} = \begin{cases} 69.02\ (T_e/n_e)^{1/2} & (SI) \\ 6.902\ (T_e/n_e)^{1/2} & (CGS) \end{cases}$$

where

	λ_D	T_e	n_e
SI	m	K	m^{-3}
CGS	cm	K	cm^{-3}

The ionic Debye length is similarly defined. When $T_i = T_e$ and $n_i = n_e$ it is

$$\lambda_{D,e} = \lambda_{D,i} = \sqrt{2}\ \lambda_D$$

The Debye length is tabulated by Drawin and Felenbok ([27] pp.427–449).

3.1.2. LANDAU LENGTH

Defined as the distance from a charge at which the electrostatic energy equals the kinetic energy,

$$l_L = K_\varepsilon \frac{e^2}{k\,T} = \begin{cases} 1.670 \times 10^{-5}\ T^{-1} & (SI) \\ 1.670 \times 10^{-3}\ T^{-1} & (CGS) \end{cases}$$

	l_L	T
SI	m	K
CGS	cm	K

3.1.3. LARMOR RADIUS

$$r_{L,\alpha} = \frac{v_\perp}{\omega_{c,\alpha}} = \begin{cases} m_\alpha\ v_\perp / Z_\alpha\ e\ B & (SI) \\ m_\alpha\ v_\perp\ c / Z_\alpha\ e\ B & (CGS) \end{cases}$$

is the radius at which a particle with charge Z_α e and mass m_α gyrates in a magnetic field B. $v_\perp$ denotes the velocity of the particle perpendicular to B, and $\omega_{c,\alpha}$ is the → cylotron frequency (see 3.3.2.). For a Maxwellian plasma with $T_i = T_e$ it follows that

$$r_{L,i} = r_{L,e}\ (m_i/m_e)^{1/2}$$

	r	m	T	B	v
SI	m	kg	K	T	m/s
CGS	cm	g	K	Gauss	cm/s

3.1.4. MEAN PARTICLE DISTANCE

$$\bar{r}_\alpha = \left(3/(4\pi\ n_\alpha) \right)^{1/3}$$

where n_α is the particle density,

	$\bar{r}_\alpha$	n_α
SI	m	m^{-3}
CGS	cm	cm^{-3}

3.1.5. THERMAL DE BROGLIE WAVELENGTH

$$\Lambda_{B,\alpha} = \frac{h}{(2\pi\ m_\alpha\ k\ T_\alpha)^{1/2}}$$

where m_α and T_α are the mass and temperature of species α.

3.1.6. MEAN-FREE-PATH

The mean free path of a test particle t is

$$\lambda_t = u / \nu_t$$

where u is the mean velocity of the test particle relative to the field particles α, and ν_t denotes the effective collision frequency with all field particles: $\nu_t = \Sigma\ \nu_{t\alpha}$. In terms of the cross-section, $\sigma_{t\alpha}$,

$$\lambda_t = \left[\sum_\alpha n_\alpha\ \sigma_{t\alpha} \left(1 + m_t/m_\alpha\right) \right]^{-1}$$

The mean-free-path of an electron in a slightly ionized plasma is

$$\lambda_e = \left[\sum n_\alpha\ \bar{\sigma}_{e\alpha} \right]^{-1}$$

and for an ion

$$\lambda_i = \frac{1}{2^{1/2}\, n_a\, \sigma_{ia} + n_e\, \sigma_{ie}(m_i/m_e)^{1/2}}$$

The mean-free-path of an electron in a completely ionized, Maxwellian plasma is

$$\lambda_{ei} = \frac{3\ (k\ T_e)^2}{2\pi\ K_\varepsilon^2\ Z^2\ e^4\ n_i\ \ell n\ \Lambda} = \begin{cases} 1.71 \times 10^9\ T_e^2\ /\ Z^2\ n_i\ \ell n\ \Lambda & (SI) \\ 1.71 \times 10^5\ T_e^2\ /\ Z^2\ n_i\ \ell n\ \Lambda & (CGS) \end{cases}$$

	λ	n_i	T_e	Z	$\ell n\ \Lambda$
SI	m	m^{-3}	K	-	-
CGS	cm	cm^{-3}	K	-	-

where Z is the mean ion charge and $\ell n\ \Lambda$ is the → Coulomb logarithm.

3.1.7. IMPACT PARAMETER FOR A 90° DEFLECTION IN A COULOMB COLLISION

$$b_{90} = \frac{K_\varepsilon\ Z_t Z_f\ e^2}{m_{tf}\ \ u^2}$$

where $Z_t e$, $Z_f e$ are the charges of the test and the field particles, $m_{tf} = m_t\ m_f/(m_t + m_f)$ the reduced mass and $u = |v_t - v_f|$ the relative velocity. For a → Maxwellian velocity distribution of test particles, the mean value is

$$\overline{b}_{90} = K_\varepsilon\ Z_t\ Z_f\ e^2\ /\ 3\ k\ T_t = \begin{cases} 5.57 \times 10^{-6}\ T^{-1} & (SI) \\ 5.57 \times 10^{-4}\ T^{-1} & (CGS) \end{cases}$$

	$\overline{b}_{90}$	T_t
SI	m	K
CGS	cm	K

3.2. CROSS-SECTIONS

3.2.1. GENERAL

The *total* collision-cross section $\sigma_{tf}(u)$ is defined as follows

$$\sigma_{tf}(u) = N_{tf} \;/\; n_t \; u$$

where N_{tf}: number of test particles colliding with the field particles per unit time

n_t : number density of test particles in the beam

u : relative velocity of test particles to field particles

$n_t u$: flux density of test particles

	σ	n	u
SI	m^2	m^{-3}	m/s
CGS	cm^2	cm^{-3}	cm/s

Particular cross-sections of a specific process (elastic: momentum transfer, inelastic: ionization, excitation, collisional recombination etc.) are defined analogously

$$\sigma_{tf}^{process} = N_{tf}^{process} / \; n_t \; u$$

where $N_{tf}^{process}$ denotes the number of test particles causing the process at a field particle per unit time.

The differential cross-section for scattering into solid angle $d\Omega = \sin\vartheta \; d\vartheta \; d\varphi$ is

$$\sigma_{tf}^{scattering}(u,\vartheta,\varphi) = N_{tf}^{scattering} \;/\; n_t \; u$$

where $N_{tf}^{(scattering)}$ denotes the number of test particles scattered by the field particle into the solid angle $d\Omega$ per unit time. Consequently,

$$\sigma_{tf}^{scattering}(u) = \int \sigma_{tf}^{scattering}(u,\vartheta,\varphi) \; d\Omega$$

If scattering may be assumed azimuthally symmetric, then

$$\sigma_{tf}^{scattering}(u) = 2\pi \int \sigma_{tf}^{scattering}(u,\vartheta) \; \sin\vartheta \; d\vartheta$$

where ϑ is the scattering angle.

3.2.2. DIFFERENTIAL RUTHERFORD CROSS-SECTION

Without exchange term in the case of identical particles,

$$\sigma^{\text{Rutherford}}(u,\vartheta) = \frac{(b_{90}/2)}{\sin^4(\vartheta/2)}$$

where b_{90} is the → impact parameter for a 90° deflection.

3.2.3. VELOCITY DEPENDENT MOMENTUM TRANSFER CROSS-SECTION

$$\sigma^{\text{momentum}}(u) = 2\pi \int_{\vartheta_{min}}^{\pi} (1 - \cos\vartheta)\, \sigma^{\text{Rutherford}}(u,\vartheta) \sin\vartheta \, d\vartheta$$

with

$$\vartheta_{min} = 2 \operatorname{arc\,ctg} \left(\lambda_{D,e}/b_{90}\right)$$

$\lambda_{D,e}$ → electronic Debye length,
b_{90} → impact parameter for a 90° deflection,

$$\sigma^{\text{momentum}}(u) = 4\pi\, b_{90}^2 \ln\left[1 + \left(\lambda_{D,e}/b_{90}\right)^2\right]^{1/2}$$

3.2.4. AVERAGE MOMENTUM TRANSFER CROSS-SECTION

$$\bar{\sigma} = \frac{2}{3} \int_0^{\infty} \mathcal{E} \exp(-\mathcal{E})\, \sigma^{\text{momentum}}\left(\left\{2\, kT\, \mathcal{E}/m_{tf}\right\}^{1/2}\right) d\mathcal{E}$$

is the energy-weighted average of the velocity dependent momentum transfer cross-section. Using $\sigma^{\text{momentum}}(u)$ for the charged particle scattering given above,

$$\bar{\sigma} = 6\pi\, \bar{b}_{90}^2 \ln \Lambda$$

when $\ln \Lambda \gg 1$, $\bar{b}_{90}$ is the → averaged value of the impact parameter for a 90° deflection. For the weaker condition, $\Lambda \gg 1$, Kihara et al. [5],[6] have shown that the equation above should be replaced by

$$\bar{\sigma} = 6\pi\, \bar{b}_{90}^2 \ln(\Lambda - 1.37)$$

Note: because $\bar{\sigma}$ does not depend on the masses of the particles involved in the collision

$$\bar{\sigma} = \bar{\sigma}_{ei} = \bar{\sigma}_{ee} = \bar{\sigma}_{ii}$$

for $Z_e = Z_i = 1$. The cross section for momentum transfer by electron-ion collisions is explicitly [7]

$$\bar{\sigma}_{ei} = \bar{\sigma}_{ie} = K_{\varepsilon} \frac{0.566}{\gamma} \frac{e^4}{(k\ T_e)^2} \ln \Lambda$$

where γ is a corretion factor of the order of 1.

3.2.5. CROSS-SECTION FOR RESONANT CHARGE EXCHANGE $A^+ + A \rightarrow A + A^+$

It is assumed that both the ion and the atom are in their ground states. Then the cross-section is given by

$$\sigma_{ia}^{ch.tr.}(E) = \sigma_0 \left[1 + a \ln\left(\frac{E_0}{E}\right)\right]^2$$

where E is the kinetic energy of the ion with respect to the atom which is at rest, $\sigma_0 = \sigma(E_0)$, $E_0 = 1$ eV. The parameters σ_0 and a (σ_0 in 10^{-19} m^2 = 10^{-15} cm^2, a dimensionless) are given in the following table:

	σ_0	a		σ_0	a		σ_0	a
H	4.9	0.15	Ti	19	0.11	Ag	17	0.11
He	2.7	0.15	V	19	0.11	Cd	14	0.11
Li	23	0.11	Cr	18	0.11	In	16	0.13
Be	11	0.12	Mn	16	0.11	Sn	8.7	0.15
B	8.3	0.15	Fe	18	0.10	Sb	9.1	0.14
C	4.3	0.18	Co	18	0.10	Te	8.6	0.14
N	3.8	0.16	Ni	16	0.11	J	5.6	0.16
O	3.5	0.18	Cu	16	0.11	Xe	7.5	0.13
F	2.5	0.18	Zn	13	0.11	Cs	45	0.09
Ne	2.5	0.17	Ga	14	0.13	Ba	30	0.10
Na	26	0.10	Ge	7.5	0.16	Ta	18	0.10
Mg	16	0.11	As	8.0	0.14	W	17	0.10
Al	13	0.14	Se	7.3	0.14	Re	19	0.10
Si	6.5	0.17	Br	4.6	0.17	Pt	13	0.11
P	7.5	0.14	Kr	5.9	0.13	Au	13	0.11
S	6.8	0.15	Rb	39	0.10	Hg	12	0.11
Cl	6.1	0.14	Sr	25	0.10	Tl	17	0.12
Ar	4.8	0.14	Zr	20	0.10	Pb	15	0.12
K	35	0.10	Nb	19	0.11	Bi	16	0.11
Ca	21	0.11	Mo	17	0.11			

3.2.6. CROSS-SECTION FOR ELECTRON IMPACT EXCITATION

The excitation process $A(l) + e^- \longrightarrow A(u) + e^-$ can be characterized by [9]

$$\sigma_{ul}^{(e)} = 4\pi\, a_0^2 \left(\frac{E_H}{E_u - E_l}\right)^2 f_{lu}\, G\left(\frac{E_e}{E_l - E_u}\right)$$

where $a_0 \longrightarrow$ Bohr radius

$E_H = 13.6$ eV ionization energy of the hydrogen atom

$E_u - E_l$: energy difference between the lower and upper levels

f_{lu}: absorption oscillator strength

$$E_e^* = \frac{E_e}{E_l - E_u}$$ translational energy of the electron in units of $E_l - E_u$

$$G\left(E_e^*\right) \cong \frac{E_e^* - 1}{E_e^*}\, \ell n\left(1.25\, \beta\, E_e^*\right)$$

$$\beta = 1 + \left(Z_{eff} - 1\right) / \left(Z_{eff} + 2\right)$$

Z_{eff}: charge seen by the bound electron (effective nuclear charge)

A similar result was obtained for optically allowed transitions at high energy (Bethe form) [196]

$$\sigma_{ul}^{(e)} = \frac{8\,\pi^2}{3^{1/2}}\, a_0^2\, \frac{E_H^2}{E_l\,|E_u - E_l|}\, f_{lu}\, \frac{g_l}{g_u}\, G_{ff}$$

where

$$G_{ff} \cong \frac{\sqrt{3}}{\pi}\, \ell n\left(4\, E_e^*\right)$$

is the free-free electron Gaunt factor.

From classical considerations, when $E_e \gg |E_u - E_l|$, the excitation cross-section my be written as [13], [196]

$$\sigma_{ul}^{(e)} = K_\varepsilon^2\, \frac{\pi\, e^4}{E_l\, |E_u - E_l|^2} \left(1 + \frac{4}{3}\frac{E_l}{E_u}\right) \frac{2\, Z^2\, E_H}{u^3}$$

3.2.7. FORMULAE FOR CROSS-SECTIONS OF ELECTRON IMPACT IONIZATION

Notation for this section:

σ_i : "total" partial cross-section for an ion i (the term "total" is used to indicate summation over the subshells n)

σ_{in} : cross-section for electron impact ionization of a target (atom or ion) leading to an ion i considering only electrons from the n-th subshell of the target

E_e : energy of the incident electron ($= m_e v_e^2/2$)

I : ionization energy of the target

I_H : ionization energy of an H-atom

I_n : binding energy of electrons in the n-th subshell $u = E_e/I$

ξ_n : number of equivalent electrons in the n-th subshell

ξ_e : number of electrons in a specific energy state for which a transfer from the incident electron of a quantity of energy, $E_1 + E_2$, would give rise to double ionization

r : mean radius of the atomic system having ξ_e electrons

Z_{eff} : (effective) nuclear charge

N : total number of electrons in the target

N_n : number of subshells

E_t : kinetic energy of the target electron

E_1 : binding energy of the first ejected electron for the case of double ionization

E_2 : binding energy of the second ejected electron in the field left by the removal of the first

....

$$\sigma_i = \sum_{n=1}^{N_n} \sigma_{in}$$

Semi-empirical approximations:

* Thomson (1912) [10]:

$$\sigma_{in} = 4\pi\, a_0^2\, \xi_n \left(\frac{I_H}{I_n}\right)^2 \frac{n-1}{n^2}$$

* Elwert (1952) [11]:

$$\sigma_{in} = 2\pi\, a_0^2\, \xi_n \left(\frac{I_H}{I_n}\right)^2 \frac{n-1}{n^2} \left[1 + 0.3\,(u-1)\right]$$

* Gryzinski (1959) [12], Stabler (1964) [13]:

$$\sigma_{in} = 4\pi\, a_0^2\, \xi_n \left(\frac{I_H}{I_n}\right)^2 \left(\frac{u}{u-1}\right)^{3/2} G(u)$$

where

$$G(u) = \begin{cases} \frac{4}{3}\frac{2^{1/2}}{u}\left(1-\frac{1}{u}\right)^{3/2} & \text{if} \quad 1 \leq u \leq 2 \\ \frac{5}{3\,u} - 2/u^2 & \text{if} \quad u > 2 \end{cases}$$

* Post (1961) [14]:

$$\sigma_{in} = 4\pi\, a_0^2\, \xi_n \left(\frac{I_H}{I_n}\right)^2 G(u)$$

where

$$G(u) = \frac{2.7183}{u}\, \ln u$$

* Drawin (1961) [15]:

$$\sigma_{in} = 2.66\pi\, a_0^2\, \xi_n \left(\frac{I_H}{I_n}\right)^2 \frac{u-1}{u^2}\, f_1\, \ln(1.25\, f_2\, u)$$

where f_1, f_2 are adjustable parameters ($\cong 1$)

* Drawin (1969) [9]:

$$\sigma_{in} = 4\pi\, a_0^2\, \xi_n \left(\frac{I_H}{I_n}\right)^2 \frac{u-1}{u^2}\, f_n\, \ln\,(1.25\, \beta_n\, u)$$

where f_n : effective oscillator strength for the ionization process $n \longrightarrow$ continuum

For the lower lying hydrogenic levels the following holds:

$f_1 \cong 0.665$; $f_2 \cong 0.71$; $f_3 = 0.81$; $f_4 \cong 0.94$; and for $n \gg 1$ $f_n \cong n/2$

$$\beta_n = 1 + \left(Z_{eff} - 1\right) / \left(Z_{eff} + 2\right)$$

* Seaton (1964) [16]:

$$\sigma_{in} = 2.2\pi\, a_0^2\, \xi_n \left(\frac{I_H}{I_n}\right)^2 \frac{u-1}{u}$$

* Gryzinski (1964) [17]:

a) single ionization

$$\sigma_{in} = 4\pi\, a_0^2\, \xi_n \left(\frac{I_H}{I_n}\right)^2 G(u)$$

where

$$g(u) = \frac{1}{u}\left(\frac{u-1}{u+1}\right)^{3/2} \left\{1 + \frac{2}{3}\left(1 - \frac{1}{2u}\right) \ln\left[2.7 + (u-1)^{1/2}\right]\right\}$$

b) double ionization (see also McFarland (1967) [18]):

$$\sigma_{in} = \left(\frac{4\pi\, a_0^2\, I_H^2}{E_1\, E_2}\right)^2 \frac{\xi_e\, (\xi_e - 1)}{\pi\, r^2}\, G\left(E_e/(E_1 + E_2)\right)$$

* Vriens (1966) [19], Burgess (1963) [20]:

$$\sigma_{in} = 4\pi\, a_0^2\, \xi_n \frac{I_H^2}{E_e + I_n + E_t}\left[\left(\frac{1}{I_n} - \frac{1}{E_e}\right) + \frac{2}{3} E_t\left(\frac{1}{I_n^2} - \frac{1}{E_e^2}\right) - \Phi\, \frac{\ln u}{E_e + I_n}\right]$$

where

$$\Phi = \begin{cases} 0 & \text{if } I_n < I_H \\ 1 & \text{if } I_n > I_H \end{cases}$$

* Lotz (1967) [21], [22], [23]:

$$\sigma_{in} = a_n\, \xi_n \frac{\ln u}{E_e\, I_n}\left[1 - b_n\, \exp\left(- c_n (u-1)\right)\right]$$

where a_n, b_n, c_n are variables

* Vainshtein, Sobelman, Yukov (1973) [24]:

$$\sigma_{in} = 4\pi\, a_0^2\, \xi_n\, (2l + 1)^{-1}\left(\frac{I_H}{I_n}\right)^2 \left(\frac{u-1}{u}\right)^{3/2} \frac{c}{u - 1 + \Phi}$$

where c and Φ are variables

* Bell et al. (1983) [25]:

$$\sigma_i(E_e) = \frac{1}{E_e I}\left\{A\,\ln u + \sum_{n=1}^{N_n} B_n \left(1 - \frac{I_n}{E_e}\right)^n\right\}$$

with A, B_n fitting parameters [25], [26]:

species	I in eV	A	B_1	B_2	B_3	B_4	B_5	reliability in %
H I	13.60	0.1845	-0.0186	0.1231	-0.1901	0.9525		± 7
He I	24.60	0.5720	-0.3440	-0.5230	3.4450	-6.8210	5.5780	± 5
He II	54.42	0.1845	0.0887	0.1315	0.3877	-1.0910	1.3541	± 10
Li I	5.39	0.0854	-0.0040	0.7573	-0.1779			± 10
Li II	75.64	0.7220	-0.1492	-1.3007	1.9443			± 12
Be I	9.32	0.9239	-0.7697	0.3619				± 20
Be II	18.21	0.7542	-0.0189	-2.9618	7.5182	-8.5431	3.1108	± 20
B I	8.30	1.1063	-1.0694	-0.0879				± 20
B II	25.15	0.9070	-0.4770	0.1970				± 20
C I	11.26	2.1143	-1.9647	-0.6084				± 5
C II	24.38	1.0824	-0.1611	-0.8563	0.9062			± 10
N I	14.53	2.2648	-1.7100	-2.3220	1.7324			± 5
N II	29.60	1.0755	-0.8287	0.8724	-0.1618	1.5331		± 10
O I	13.62	2.4554	-2.1811	-1.5701				± 5
O II	35.12	1.5257	-0.5935	-0.3994	-0.5833	3.2355		± 10

Remarks:

(i) Thomson's model (classical binary-encounter approximation, target is at rest) gives too sharp a peak in the cross-section at too low an energy so that the cross-section decreases too rapidly at high electron energy where the correct behaviour is $\sigma \propto (\ln T)/T$ rather than $\sigma \propto 1/T$.

(ii) The most widely used empirical formulae are due to Lotz [21], [22], [23]. For ions with charge ≥ 3, Lotz approximates the "total" cross- section by

$$\sigma_i = \sum_{n=1}^{N_n} \xi_n \frac{a\,\ln u}{E_e\,I_n} \qquad \text{with} \qquad a = 4.5 \times 10^{-14}\ \mathrm{cm}^2\ (\mathrm{eV})^2$$

(iii) Seaton's formula should be applied only to near threshold collisions and for highly charged ions.

References: [26]-[29]

3.2.8. CROSS-SECTION OF PHOTON SCATTERING BY FREE ELECTRONS

$$\sigma^{ph.sc.} = \frac{8\pi}{3}\left(\frac{K_\varepsilon e^2}{m_e c^2}\right)^2 = \frac{8\pi}{3} r_e^2 = \begin{cases} 6.653 \times 10^{-29}\ m^2 & (SI) \\ 6.653 \times 10^{-25}\ cm^2 & (CGS) \end{cases}$$

is valid in the non-relativistic case, i.e.

$$h\nu \ll m_e c^2$$

or

$$\nu \ll 1.24 \times 10^{20}\ s^{-1}$$

or

$$\lambda \gg 2.43 \times 10^{-12}\ m = 0.0243\ \text{Å}$$

r_e is the → classical electron radius.

References: [30], [31]

3.2.9. CROSS-SECTION FOR PHOTON ABSORPTION

The cross section for a bound-bound absorption corresponding to an atomic transition from the lower level, *l*, to the upper level, *u*, is [32]

$$\sigma_\nu^{(l\to u)} = \frac{K_\varepsilon \pi e^2 f_{lu} P(\nu)}{m_e c} = \begin{cases} 2.654 \times 10^{-6} f_{lu} P(\nu) & (SI) \\ 2.654 \times 10^{-2} f_{lu} P(\nu) & (CGS) \end{cases}$$

where f_{lu} denotes the → absorption oscillator strength,

$P(\nu)$ absorption → line-shape factor, $\int_0^\infty P(\nu)\, d\nu = 1$,

	σ	f_{lu}	$P(\nu)$
SI	m^2	–	s
CGS	cm^2	–	s

3.2.10. CROSS-SECTION OF PHOTO-RECOMBINATION

The semi-classical formula for the cross-section of radiative capture to a hydrogenic level with principal quantum number *l*, first calculated by Kramers, 1923, has the form [32]

$$\sigma_{e\rightarrow l}^{ph.rec.}(v_e) = K_\varepsilon^5 \frac{2^7\ \pi^4\ \ Z_{eff}^4\ e^{10}}{3\ \ 2^{1/2}\ m_e\ v_e^2\ c^3\ h^4\ \nu\ l^3}\ G\left(\frac{2\pi\ \nu\ b_0}{v_e}\right)$$

$$= \begin{cases} 3.32 \times 10^2\ \ Z_{eff}^4\ G\ /\ v_e^2\ \nu\ l^3 & (SI) \\ 3.32 \times 10^{10}\ Z_{eff}^4\ G\ /\ v_e^2\ \nu\ l^3 & (CGS) \end{cases}$$

	$\sigma_{e\rightarrow l}^{ph.rec.}$	ν	Z_{eff}	l	v_e	G
SI	m^2	s^{-1}	-	-	m/s	-
CGS	cm^2	s^{-1}	-	-	cm/s	-

where Z_{eff}: effective nuclear charge

ν : frequency of the emitted radiation ($h\nu$ photon energy)

b_0 : → impact parameter for 90° scattering

v_e : velocity of the electron before being trapped

The correction function $G\left(\frac{2\pi\ \nu\ b_0}{v_e}\right)$ differs from unity by less than 20% for $(2\pi\ \nu\ b_0/v_e) > 0.2$. Radiation at low frequencies corresponds to collisions with large impact parameters b. The effects of plasma shielding would limit collisions to impact parameters of the order or less than the → Debye length, and there would thus be a cut-off in radiation for frequencies of the order $2\pi\ \nu_{min} \cong v_e\ /\lambda_D$.

The quantum theoretical result has the same form as the semi-classical relation due to Kramers except that the bound-free Gaunt factor G_{bf} replaces the function $G(2\pi\ \nu\ b_0/v_e)$. In most cases of interest, the value of G_{bf} is very close to unity.

The frequency of the emitted radiation is given by

$$\nu = \frac{1}{h}\left(\frac{m_e\ v_e}{2} + E_{l\rightarrow e}\right)^2$$

where $E_{l\rightarrow e}$ denotes the ionization energy of the newly formed particle. Introducing this expression into Kramers' formula and using hydrogenic energy levels, we obtain

$$\sigma^{ph.rec.}_{e\rightarrow l}(v_e) = \begin{cases} 1.01 \times 10^{-13} \dfrac{Z_{eff} \; G_{bf}}{v_e^2 \, l \left[1 + (m_e \; v_e^2/2 \; E_{l\rightarrow e}) \right]} & (SI) \\ 1.01 \times 10^{-5} \dfrac{Z_{eff} \; G_{bf}}{v_e^2 \, l \left[1 + (m_e v_e^2/2 \; E_{l\rightarrow e}) \right]} & (CGS) \end{cases}$$

Other versions are given by Biberman [33]

$$\sigma^{ph.rec.}_{e\rightarrow l}(E_e) \cong \pi \, a_0^2 \, \frac{32}{3\sqrt{3}} \, \frac{Z^4_{eff} \; (Ry)^2_E \; \alpha^3_F}{h\nu \;\; l^3 \;\; E_e}$$

and by Spitzer [31]

$$\sigma^{ph.rec.}_{e\rightarrow l}(E_e) = A_{ph.rec.} \; \frac{\nu_0}{\nu} \; \frac{h\nu_0}{E_e} \; \frac{g_{e\rightarrow l}}{l^3}$$

where

$$A_{ph.rec.} = K_\varepsilon \; \frac{2^4}{3^{3/2}} \; \frac{h \; e^2}{\pi \; m_e^2 \; c^3} = \begin{cases} 2.106 \times 10^{-26} \; m^2 & (SI) \\ 2.106 \times 10^{-22} \; cm^2 & (CGS) \end{cases}$$

is the constant of electron capture

E_e : kinetic energy of the electron before being trapped ($= m_e v_e^2/2$)

$h\nu_0$: energy of the ground state $\left(= Z^2_{eff} (Ry)_E \right)$

α_F : fine-structure constant (= 1/137)

$G_{e\rightarrow l}$ Gaunt factor of the transition: free $\rightarrow$ level with principal quantum number l; For electron capture into the ground state

$$G_{e\rightarrow 1} = 8\pi \sqrt{3} \; \frac{\nu_0}{\nu} \; \frac{exp\left\{-4 \left[\nu_0/(\nu - \nu_0)\right]^{1/2} arctg\left[(\nu - \nu_0)/\nu_0\right]^{1/2}\right\}}{1 - exp\left\{-2\pi \left[\nu_0/(\nu - \nu_0)\right]^{1/2}\right\}}$$

$G_{e\rightarrow l} = 0.797$ if $\nu \cong \nu_0$, i.e. the electron captured was of low energy, $E_e \ll h\nu_0$. For estimations $g_{e\rightarrow l} \cong 1$ is sufficient.

References: [31], [32]

3.3. CHARACTERISTIC TIMES AND FREQUENCIES

3.3.1. PLASMA FREQUENCY (LANGMUIR FREQUENCY)

$$\omega_{pl} = \left\{ K_\varepsilon \, 4\pi \, e^2 \left[\frac{n_e}{m_e} + \sum_Z \left(\frac{Z^2 \, n_i^{z+}}{m_i} \right) \right] \right\}^{1/2}$$

where Z is the ion charge and n_i^{z+} is the density of Z-fold charged ions. Often, the electron plasma frequency

$$\omega_{pl,e} = \left(\frac{K_\varepsilon \, 4\pi \, e^2 \, n_e}{m_e} \right)^{1/2} = \begin{cases} 56.4 \; n_e^{1/2} & (SI) \\ 5.64 \times 10^4 \; n_e^{1/2} & (CGS) \end{cases}$$

is used.

Analogously, the ion plasma frequency is defined as

$$\omega_{pl,i} = \left(\frac{K_\varepsilon \, 4\pi \, e^2}{m_i} \sum_Z Z^2 \, n_i^{z+} \right)^{1/2}$$

	ω_{pl}	n	m
SI	s^{-1}	m^{-3}	kg
CGS	s^{-1}	cm^{-3}	g

Note, sometimes the expressions

$$f_{pl,e} = \omega_{pl,e}/2\pi \qquad \text{and} \qquad f_{pl,i} = \omega_{pl,i}/2\pi$$

are also used for the electron and ion plasma frequencies.

3.3.2. GYRATION FREQUENCY (CYCLOTRON FREQUENCY, LARMOR FREQUENCY)

A particle α having charge $Z_\alpha e$ and mass m_α gyrates in a magnetic field B with the frequency

$$\omega_{c,\alpha} = 2\pi \, f_{c,\alpha} = \begin{cases} Z_\alpha \, e \, B \, / \, m_\alpha & (SI) \\ Z_\alpha \, e \, B \, / (m_\alpha \, c) & (CGS) \end{cases}$$

For electrons:

$$\omega_{c,e} = \begin{cases} 1.76 \times 10^{11}\ B & (SI) \\ 1.76 \times 10^{7}\ B & (CGS) \end{cases}$$

	f,ω	B	m_α	M_α	Z_α
SI	s^{-1}	T	kg	-	-
CGS	s^{-1}	G	g	-	-

For ions:

$$\omega_{c,i} = \begin{cases} 9.55 \times 10^{7}\ Z_\alpha\ B\ /\ M_\alpha & (SI) \\ 9.55 \times 10^{3}\ Z_\alpha\ B\ /\ M_\alpha & (CGS) \end{cases}$$

where $M_\alpha = m_\alpha/m_u$, m_u denotes the → atomic mass unit.

3.3.3. HYBRID FREQUENCIES

(i) lower hybrid frequency

$$\omega_{LH} = \left(\omega_{c,i}^2 + \frac{\omega_{c,e}^2\ \omega_{pl,i}^2}{\omega_{c,e}^2 + \omega_{pl,e}^2} \right)^{1/2}$$

(ii) upper hybrid frequency

$$\omega_{UH} = \left[\omega_{pl}^2 + (\omega_{c,e})^2 \right]^{1/2}$$

3.3.4. COLLISION FREQUENCIES

(i) General

The frequency at which a single test particle t collides with field particles f of a gas (plasma) is given by

$$\nu_{tf}(u) = \sum_f n_f\ u\ \sigma_{tf}(u)$$

where

u : relative velocity of the test and field particles

$\sigma_{tf}(u)$: collision cross-section (if σ represents a total cross-section, ν will be a total collision frequency. If σ corresponds to a particular process, ν will be the collision frequency of that process)

n_f : density of field particles

(ii) Average momentum transfer collision frequencies

The effective (average) frequency of collisions of electrons with heavy particles f is given by [32], [36]

$$\nu_e = \sum_f \nu_{ef}$$

where

$$\nu_{ef} = -\frac{4\pi}{3}\int_0^\infty n_f\, \sigma_{ef}(u)\, u^4 \frac{\partial f_e}{\partial u}\, du$$

with $\sigma_{ef}(u)$: cross-section for momentum transfer

$f_e(u)$: electron distribution function

For electron-ion momentum transfer in a plasma, the screened Rutherford cross-section is used. With a → Maxwellian electron distribution this gives

$$\nu_{ei} = K_\varepsilon^2 \frac{4\,(2\pi)^{1/2}\, Z^2\, e^4\, n_i\, \ell n\, \Lambda}{3\,(m_e)^{1/2}\,(kT_e)^{3/2}}$$

$$= \begin{cases} 3.63 \times 10^{-6}\, Z^2\, n_i\, \ell n\, \Lambda\, /T_e^{3/2} & (SI) \\ 3.63\; Z^2\, n_i\, \ell n\, \Lambda\, /\, T_e^{3/2} & (CGS) \end{cases}$$

	ν_{ei}	n_i	$\ell n\, \Lambda$	T_e
SI	s^{-1}	m^{-3}	-	K
CGS	s^{-1}	cm^{-3}	-	K

Similarly, the electron-electron momentum transfer collision frequency is

$$\nu_{ee} = K_\varepsilon^2 \frac{4\,\pi^{1/2}\, e^4\, n_e\, \ell n\, \Lambda}{3\, m_e^{1/2}\,(kT_e)^{3/2}}$$

$$= \begin{cases} 2.57 \times 10^{-8}\, n_e\, \ell n\, \Lambda\, /\, T_e^{3/2} & (SI) \\ 2.57 \times 10^{-2}\, n_e\, \ell n\, \Lambda\, /\, T_e^{3/2} & (CGS) \end{cases}$$

and the ion-ion momentum transfer collision frequency

$$\nu_{ii} = K_\varepsilon^2 \frac{4\,\pi^{1/2}\, Z^4\, e^4\, n_i\, \ell n\, \Lambda}{3\, m_i^{1/2}\,(kT_i)^{3/2}}$$

$$\nu_{ii} = \begin{cases} 6.01 \times 10^{-8}\ Z^4\ n_i\ \ln\Lambda\ /\ T_i^{3/2}\ M_i^{1/2} & (SI) \\ 6.01 \times 10^{-2}\ Z^4\ n_i\ \ln\Lambda\ /\ T_i^{3/2}\ M_i^{1/2} & (CGS) \end{cases}$$

where M_i is the ion mass in → atomic mass units.

If $n_e = n_i$ and $T_e = T_i$ then $\nu_{ii} = \nu_{ee} \left(m_e/m_i\right)^{1/2}$.

(iii) Energy exchange collision frequency

In the case of different masses of particles involved in the collision, the energy transfer can be weak. The energy exchange collision frequency of electrons is

$$\nu_{eh}^{E} = \frac{2m_e}{m_h}\ \nu_{eh}^{m.tr.}$$

where $\left(\nu_{eh}^{E}\right)^{-1}$ represents a characteristic time in which an electron is able to give its energy to heavy particles by elastic collisions, $\nu_{eh}^{m.tr.}$ is the corresponding collision frequency of momentum transfer. The average fraction of the kinetic energy lost by an electron in an elastic collision with a heavy particle is

$$\Delta E \cong \frac{2m_e}{m_h}\ \overline{E} \qquad \text{with} \qquad \overline{E} = \frac{1}{2}\ m_e\ v_e^2 = \frac{1}{2}\ m_e \left(\frac{8\ kT_e}{\pi\ m_e}\right)$$

The last equality is true in the case of a Maxwellian electron distribution. Then

$$\Delta E \cong \frac{8\ kT_e}{\pi}\ \frac{m_e}{m_h}$$

(iv) Frequency of a 90° deflection in Coulomb collisions

For electrons moving through the field of nearly motionless ions, a 90° deflection frequency can be defined by

$$f_{90} = 8\pi\ n_i\ u\ b_{90}^2\ \ln\Lambda = K_\varepsilon^2\ 8\pi\ \frac{n_i\ Z_i^2\ e^4\ \ln\Lambda}{m_i^2\ u^3}$$

$$= \begin{cases} 4.85 \times 10^{-1}\ n_i\ Z_i^2\ \ln\Lambda\ /\ M_i^2\ u^3 & (SI) \\ 4.85 \times 10^{11}\ n_i\ Z_i^2\ \ln\Lambda\ /\ M_i^2\ u^3 & (CGS) \end{cases}$$

where

b_{90} : → impact parameter for a 90° deflection (see 3.1.7.)

n_i : ion density

u : relative ion-electron velocity

$\ell n\ \Lambda$: → Coulomb logarithm

M_i : ion mass in → a.m.u.

	f_{90}	b_{90}	n_i	Z_i	$\ell n\ \Lambda$	m_i	M_i	u
SI	s^{-1}	m	m^{-3}	-	-	kg	a.m.u	m/s
CGS	s^{-1}	cm	cm^{-3}	-	-	g	a.m.u.	cm/s

Reference: [31] p.77

3.3.5. MAXWELLIZATION TIME FOR AN ENSEMBLE OF INTERACTING CHARGED PARTICLES

This is given by

$$\tau_\alpha^{maxwell} = \frac{m_\alpha^{1/2}\ (3\ kT_\alpha)^{3/2}}{K_\varepsilon^2\ \ 0.714\ \ 8\pi\ Z_\alpha^4\ e^4\ n_\alpha\ \ell n\ \Lambda}$$

where m_α, T_α, n_α, Z_α denotes respectively the mass, temperature, number density, and mean charge of species α.

For electrons, $\alpha = e$, $Z_e = 1$

$$\tau_e^{maxwell} \cong \nu_{ee}^{-1}$$

that is, the inverse momentum transfer collision frequency. The factor 0.714 is the result of

$$\Phi\left[\left(\frac{m_\alpha}{2\ kT_\alpha}\right)^{1/2} v_\alpha\right] - G\left[\left(\frac{m_\alpha}{2\ kT_\alpha}\right)^{1/2} v_\alpha\right]$$

with

$$\left(\frac{m_\alpha}{2\ kT_\alpha}\right)^{1/2} v_\alpha = \left(\frac{3}{2}\right)^{1/2}$$

and

$$\Phi(x) = \frac{2}{\pi^{1/2}} \int_0^x exp\left(-y^2\right) dy \quad \text{and} \quad G(x) = \frac{1}{2\ x^2}\left(\Phi(x) - x\,\frac{d\Phi}{dx}\right)$$

Reference: [31]

3.3.6. TEMPERATURE EQUILIBRATION TIME OF TWO INTERACTING ENSEMBLES OF CHARGED PARTICLES

Assuming each ensemble to have a → Maxwellian distribution with a temperatures T_α and T_β, respectively, then

$$\frac{dT_\alpha}{dt} = \frac{T_\alpha - T_\beta}{\tau^{equal}}$$

$$\tau^{equal} = \frac{3\ m_\alpha\ m_\beta}{K_\varepsilon^2\ 8\ (2\pi)^{1/2}\ n_\beta\ Z_\alpha^2\ Z_\beta^2\ e^4\ \ell n\ \Lambda} \left(\frac{kT_\alpha}{m_\alpha} + \frac{kT_\beta}{m_\beta} \right)^{3/2}$$

Reference: [31]

3.3.7. THERMALIZATION TIME OF A FAST PARTICLE

Assuming that a fast, charged, test particle t collides with field particles f having a Maxwellian distribution at a temperature T_f, then

$$\tau^{thermal} = \frac{v_t\ m_t^2\ kT_f}{K_\varepsilon^2\ 4\pi \left(1 + \frac{m_t}{m_f}\right) m_f\ Z_t^2\ Z_f^2\ e^4\ n_f\ \ell n\ \Lambda\ \ G\left[\left(\frac{m_f}{kT_f}\right)^{1/2} v_t\right]}$$

The function G(x) has already been introduced (→ Maxwellization time).

a) If $v_t \gg \left(3\ kT_f/m_f\right)^{1/2}$ then

$$\tau^{thermal} = \frac{v_t^3\ m_t^2}{K_\varepsilon^2\ 4\pi \left(1 + \frac{m_t}{m_f}\right) Z_t^2\ Z_f^2\ e^4\ n_f\ \ell n\ \Lambda}$$

b) If $v_t \ll \left(3\ kT_f/m_f\right)^{1/2}$ then

$$\tau^{thermal} = \frac{3\ (2\pi)^{1/2}\ m_t\ \ (kT_f)^{3/2}}{K_\varepsilon^2\ 8 \left(1 + \frac{m_t}{m_f}\right) m_f^{3/2}\ Z_t^2\ Z_f^2\ e^4\ n_f\ \ell n\ \Lambda}$$

Reference: [31]

3.4. VELOCITIES

3.4.1. GENERAL

The <u>species average velocity</u> relative to the laboratory frame is defined as

$$v_\alpha = \int_{-\infty}^{\infty} v' \, f_\alpha(v') \, d^3v'$$

where $f_\alpha(v')$ is the velocity distribution function of species α. The <u>mean mass velocity of the fluid (plasma) as a whole</u> is defined by

$$v(r,t) = \left(\sum \rho_\alpha v_\alpha\right) \Big/ \left(\sum \rho_\alpha\right)$$

where $\rho_\alpha(r,t) = m_\alpha \; n_\alpha(r,t)$ denotes the mass density of particles α. The <u>diffusion velocity</u> of a species α is defined by the difference

$$u_\alpha(r,t) = v_\alpha(r,t) - v(r,t)$$

The diffusion velocities must satisfy the relation $\sum_\alpha \rho_\alpha u_a = 0$.

It is convenient to consider particle velocities V' with respect to a local frame of reference moving with the mean mass velocity v (*peculiar velocity*) V'= v'- v. Then, the <u>diffusion velocity</u> of species α can be written as

$$U_\alpha(r,t) = \int_{-\infty}^{\infty} V' \, f_\alpha(V') \, d^3V'$$

3.4.2. THERMAL VELOCITY

The mean velocity of particles α with a → Maxwellian velocity distribu- tion is given by

$$\bar{v}_\alpha = \left(\frac{8 \; kT_\alpha}{\pi \; m_\alpha}\right)^{1/2}$$

Often, the expression

$$\bar{v}_{th,\alpha} = \left(\frac{kT_\alpha}{m_\alpha}\right)^{1/2}$$

is refered to as the thermal velocity. The latter is, for electrons,

$$\bar{v}_{th,e} = \begin{cases} 3.893 \times 10^{3}\ T_e^{1/2} & (SI) \\ 3.893 \times 10^{5}\ T_e^{1/2} & (CGS) \end{cases}$$

with

	v_{th}	T_e
SI	m/s	K
CGS	cm/s	K

or

$$\bar{v}_{th,e} = \begin{cases} 4.194 \times 10^{5}\ T_e^{1/2} & (SI) \\ 4.194 \times 10^{7}\ T_e^{1/2} & (CGS) \end{cases}$$

with

	v_{th}	T_e
SI	m/s	eV
CGS	cm/s	eV

3.4.3. ALFVEN VELOCITY

The phase velocity of magnetic waves (Alfvén waves) is

$$v_A = \begin{cases} \dfrac{B}{[\mu_0\ (n_i m_i + n_e m_e)]^{1/2}} \cong \dfrac{B}{[\mu_0\ n_i m_i]^{1/2}} = 2.19 \times 10^{16}\ B/(M_i n_i)^{1/2} & (SI) \\ \dfrac{B}{[4\pi\ (n_i m_i + n_e m_e)]^{1/2}} \cong \dfrac{B}{[4\pi\ n_i m_i]^{1/2}} = 2.19 \times 10^{11}\ B/(M_i n_i)^{1/2} & (CGS) \end{cases}$$

	v	B	n	m	M
SI	m/s	T	m^{-3}	kg	-
CGS	cm/s	G	cm^{-3}	g	-

where B denotes the magnetic induction (magnetic field) and M_i is the atomic weight of the ions.

Reference: [37]

3.4.4. SOUND VELOCITY

The general definition is

$$v_s^2 = \left(\frac{\partial p}{\partial \rho}\right)_{s,c}$$

where p, ρ denote the pressure and mass density respectively. The subscripts s,c indicate that the entropy and the composition of the gas (plasma) are held constant during the differentiation. With the equation of state of an ideal gas it follows that

$$v_{s,g} = \left(\varkappa\ R_g\ T\right)^{1/2}$$

where $\varkappa \longrightarrow$ ratio of specific heats

$R_g = R/m_g$ specific gas constant (ratio of the → universal gas constant to the mass of the gas particle)

T : temperature of the gas

When the gas is essentially ionized, the equation above must be modified to

$$v_s = \left(\frac{\varkappa R_g T}{1 - (\alpha/2)} \right)^{1/2}$$

where α is the → degree of ionization, $\alpha = n_i/(n_i + n_a)$.

In a plasma, there are two different modes of longitudinal waves. The first mode represents an ion wave. At low frequencies, the speed of propagation is the sound speed of the plasma as a whole (see above). At very high frequencies, it is equal to the ion sound speed

$$v_{s,i} = \left(\frac{\varkappa p_i}{m_i n_i} \right)^{1/2}$$

where p_i, m_i, n_i denote the partial pressure, mass, and number density of ions. The velocity of ion sound waves ("ion sound velocity") may be expressed as

$$v_s = \left(\frac{\varkappa_e kT_e + \varkappa_i kT_i}{m_i} \right)^{1/2}$$

For plane waves, $\varkappa_e = 1$, $\varkappa_i = 3$. If $T_i \ll T_e$

$$v_s \approx \left(kT_e / m_i \right)^{1/2}$$

The second mode represents an electron wave.

At low frequencies, $\omega < \omega_{pl,e}$, the wave is damped exponentially whilst at high frequencies, $\omega > \omega_{pl,e}$, it is an undamped wave the speed of which is the electron sound speed

$$v_{s,e} = \left(\frac{\varkappa p_e}{m_e n_e} \right)^{1/2}$$

In magnetogasdynamics, the ratio of the magnetic pressure to the total pressure should be taken into account, leading to the effective sound speed of the gas (plasma):

$$v^2_{s,eff} = \frac{\partial}{\partial \rho} \left(p + \frac{B}{2 \mu_0} \right) = \frac{\varkappa p}{\rho} + \frac{2 p_m}{\rho}$$

$$v^2_{s,eff} = v^2_s + v^2_A$$

$$= \varkappa R_g T + B^2/(\rho \mu_0)$$

where μ_0 : → permeability of vacuum

p : gas dynamic pressure

$p_m = B^2/(2 \mu_0)$: magnetic pressure

ρ : mass density

v_s : (ordinary) sound velocity

v_A : → Alfvén velocity

References: [37] vol.3 p.63; [38] pp.107, 163

3.5. DIMENSIONLESS VARIABLES

3.5.1. DEGREE OF IONIZATION

(1) Degree of ionization α

$$\alpha = \frac{n_e}{n_a + n_e}$$

(2) Degree of ionzation x

$$x = \frac{n_e}{n_a + n_i}$$

where

n_e : number density of free electrons

n_a : number density of (neutral) atoms

n_i : number density of ions; $n_a + n_i$ is the number density of heavy particles

If only singly charged ions occur in any plasma, both these definitions are identical. If multiply charged ions are present, the quantities α and x satisfy the following equalities

$$0 \le \alpha \le 1 \qquad \text{and} \qquad 0 \le x < \infty$$

3.5.2. COULOMB LOGARITHM

The Coulomb logarithm originates from an integration over the impact parameter b for Coulomb collisions. Cut-offs due to physical processes are uncertain to within a numerical factor of order unity. The Coulomb logarithm is defined for a low temperature plasma by ([31] p.170, [33] p.20, [41] p.7)

$$\ell n\, \Lambda = \ell n \left(\frac{\lambda_{D,e}}{\bar{b}_{90^\circ}} \right) = \ell n \left(3 \frac{\lambda_{D,e}}{l_L} \right)$$

$$= \ell n \left(K_\varepsilon^{-3/2} \frac{3}{2} \frac{(kT)^{3/2}}{\pi^{1/2}\, e^3\, n_e^{1/2}} \right)$$

$$= \begin{cases} \ell n \left(1.24 \times 10^7\, T^{3/2} / n_e^{1/2} \right) & (SI) \\ \ell n \left(1.24 \times 10^4\, T^{3/2} / n_e^{1/2} \right) & (CGS) \end{cases}$$

	Λ	T	n_e
SI	-	K	m^{-3}
CGS	-	K	cm^{-3}

$$= \begin{cases} \ell n \left(1.55 \times 10^{13}\, T^{3/2} / n_e^{1/2} \right) & (SI) \\ \ell n \left(1.55 \times 10^{10}\, T^{3/2} / n_e^{1/2} \right) & (CGS) \end{cases}$$

	Λ	T	n_e
SI	-	eV	m^{-3}
CGS	-	eV	cm^{-3}

where

$\bar{b}_{90^\circ}$ → mean impact parameter for a 90° deflection in a Coulomb collision
$\lambda_{D,e}$ → electronic Debye length
l_L → Landau length

Refined calculations give asymptotic expressions for $\ell n\, \Lambda$ for the various limiting cases of free-free transitions. They shoold be used in the expressions for the effective electron-ion collision frequency where a Maxwellian distribution of test particles may be assumed. These expressions are [42] p.87

	low frequency, $\omega \ll \omega_{pl}$ (high density, $n_e \gg n_{cr}$)	high frequency, $\omega \gg \omega_{pl}$ (low density, $n_e \ll n_{cr}$)
low temperature $kT \ll Z^2$ Ry	$\ln\left[\frac{1}{3\,e^{1/2}}\left(\frac{2}{\gamma}\right)^2 \Lambda_{sp}\right]$ $= \ln\left[0.255\,\Lambda_{sp}\right]$	$\ln\left[\frac{1}{3}\left(\frac{2}{\gamma}\right)^{5/2}\frac{\omega_{pl}}{\omega}\Lambda_{sp}\right]$ $= \ln\left[0.445\,\frac{\omega_{pl}}{\omega}\Lambda_{sp}\right]$
high temperature $kT \gg Z^2$ Ry	$\ln\left[\left(\frac{8}{e\gamma}\right)^{1/2}\frac{kT}{\hbar\,\omega_{pl}}\right]$ $= \ln\left[1.285\,\frac{kT}{\hbar\,\omega_{pl}}\right]$	$\ln\left[\frac{4}{\gamma}\,\frac{kT}{\hbar\,\omega}\right]$ $= \ln\left[2.246\,\frac{kT}{\hbar\,\omega}\right]$

where

$\Lambda_{sp} = \lambda_{D,e}/\overline{b}_{90°}$ is the expression defined at the beginning of this section

ω : frequency of the emitted photon

ω_{pl} : → plasma frequency

n_{cr} : critical density

$$n_{cr} = K_\varepsilon^{-1}\,\frac{m_e}{4\pi\,e^2}\,\omega^2 = \begin{cases} 3.14 \times 10^{-4}\,\omega^2 & (SI) \\ 3.14 \times 10^{-10}\,\omega^2 & (CGS) \end{cases}$$

	n_{cr}	ω
SI	m^{-3}	s^{-1}
CGS	cm^{-3}	s^{-1}

Ry → Rydberg constant in energy units (13.60 eV)

γ = 1.781... Euler's constant

e = 2.718...

$\hbar = h/2\pi$ → Planck's constant

Considering collisions between various species in a plasma, the following expressions can be used over a wide range of plasma parameters [35] (n_e in cm^{-3}, T_e in eV):

Thermal electron-electron collisions

$$\ell n\,\Lambda = \begin{cases} 23 - \ell n\left(n_e^{1/2}\,T_e^{-3/2}\right) & \text{for } T_e \le 10\text{ eV} \\ 24 - \ell n\left(n_e^{1/2}\,T_e^{-1}\right) & \text{for } T_e \ge 10\text{ eV} \end{cases}$$

Electron-ion collisions

$$\ell n\,\Lambda = \begin{cases} 23 - \ell n\left(n_e^{1/2}\,Z\,T_e^{-3/2}\right) & \text{for } Z^2\,10\text{ eV} > T_e > T_i\,m_e/m_i \\ 24 - \ell n\left(n_e^{1/2}\,T_e^{-1}\right) & \text{for } T_e > Z^2\,10\text{ eV} > T_i\,m_e/m_i \\ 30 - \ell n\left(n_i^{1/2}\,T_i^{-3/2}\,Z^2\,M_i^{-1}\right) & \text{for } T_e < T_i\,Z\,m_e/m_i \end{cases}$$

Mixed ion-ion collisions

$$\ell n\,\Lambda = 23 - \ell n\left\{\frac{Z_\alpha\,Z_\beta\,(M_\alpha + M_\beta)}{M_\alpha\,T_{i,\beta} + M_\beta\,T_{i,\alpha}}\left(\frac{n_{i,\alpha}\,Z_\alpha^2}{T_{i,\alpha}} + \frac{n_{i,\beta}\,Z_\beta^2}{T_{i,\beta}}\right)^{1/2}\right\}$$

Counterstreaming ions in the presence of warm electrons
(relative ion velocity v_D; $kT_{i,\alpha}/m_{i,\alpha}$, $kT_{i,\beta}/m_{i,\beta} < v_D < kT_e/m_e$)

$$\ell n\,\Lambda = 35 - \ell n\left\{\frac{Z_\alpha\,Z_\beta\,(M_\alpha + M_\beta)}{M_\alpha\,M_\beta\,(v_D/c)^2}\left(\frac{n_e}{T_e}\right)^{2}\right\}$$

with

$M_\alpha = m_{i,\alpha}/m_u$: atomic mass number of ions of the kind α
Z_α, $n_{i,\alpha}$, $T_{i,\alpha}$: their charge, density, and temperature

References: [31], [33], [35], [41]-[44]

3.5.3. PARAMETERS OF STRONGLY COUPLED PLASMAS

A plasma is said to be ideal if the mean interaction energy of particles is much less than their kinetic energy. With respect to the Coulomb interaction and the particle thermal energy, the parameter for a strongly coupled plasma γ can be defined by ([40] pp.6,11, [41], [45] p.41)

$$\gamma = K_\varepsilon \frac{e^2 (n_e + n_i)^{1/3}}{k\,T} = \begin{cases} 1.67 \times 10^{-5} (n_e + n_i)^{1/3} / T & (SI) \\ 1.67 \times 10^{-3} (n_e + n_i)^{1/3} / T & (CGS) \end{cases}$$

	n	T	γ
SI	m^{-3}	K	-
CGS	cm^{-3}	K	-

If $\gamma \ll 1$ the plasma is said to be ideal, whilst if $\gamma \to 1$ the plasma is said to be non-ideal (strongly coupled plasma, dense plasma, non-Debye plasma). In the literature, a definition is also used in which $n_e + n_i$ is replaced by n_e. These definitions are applicable to classical particle systems where $E_F/kT \ll 1$ is valid (E_F denotes the → Fermi energy).

In degenerated plasmas, kT should be replaced by E_F. The criterion for a non-ideal plasma then reads

$$\gamma^{quant} = K_\varepsilon \frac{e^2\, n_e^{1/3}}{E_F}$$

Because of the different electron and ion masses, different parameters describing a non-ideal plasma were introduced (Brueckner's parameter)

$$\gamma_e^{quant} = \bar{l}_e/a_0 = \left(3/(4\pi\, n_e)\right)^{1/3} \Big/ \left(\hbar^2/(K_\varepsilon\, e^2\, m_e)\right)$$

$$= \begin{cases} 1.17 \times 10^{10}\, n_e^{-1/3} & (SI) \\ 1.17 \times 10^{8}\, n_e^{-1/3} & (CGS) \end{cases}$$

$$\gamma_i^{quant} = \left(3/(4\pi\, n_i)\right)^{1/3} \Big/ \left(\hbar^2/(K_\varepsilon\, e^2\, m_i)\right)$$

$$= \begin{cases} 2.15 \times 10^{13}\, n_i^{-1/3}\, M_i & (SI) \\ 2.15 \times 10^{11}\, n_i^{-1/3}\, M_i & (CGS) \end{cases}$$

	γ^{quant}	$n_{e/i}$	M_i
SI	-	m^{-3}	-
CGS	-	cm^{-3}	-

where $\bar{l}_e$ denotes the mean electron distance, and M_i is the atomic mass number of the ions.

In place of γ, alternative parameters can be used to describe a non-ideal plasma. These are

<u>(i) the number of charged particles in the Debye sphere</u>

$$N_D = \frac{4}{3} \pi \, (n_e + n_i) \, \lambda_D^3$$

<u>(ii) the ratio of the → Debye length and the → mean impact parameter for a 90° deflection in an electron-ion collision</u>
(= ratio of the → Debye length and 1/3 of the → Landau length, see 3.5.2.)

$$\Lambda = \frac{3 \, \lambda_D \, kT}{K_\varepsilon \, e^2} = \frac{3 \, kT^{3/2}}{K_\varepsilon^{3/2} \, 2 \, \pi^{1/2} \, e^3 \, n_e^{1/2}}$$

The r.h.s. follows from using the <u>electronic</u> Debye radius (e.g. [31]), but the complete Debye length is sometimes used. In this case the expression given should be multiplied by the factor $2^{-1/2}$ ([40], [41]).

<u>(iii) plasma parameter</u>

$$\Gamma = K_\varepsilon \frac{e^2}{\lambda_{D,e} \, kT} = \begin{cases} 2.42 \times 10^{-7} \, (n_e/T^3)^{1/2} & (SI) \\ 2.42 \times 10^{-4} \, (n_e/T^3)^{1/2} & (CGS) \end{cases}$$

	n_e	T	Γ
SI	m^{-3}	K	-
CGS	cm^{-3}	K	-

The relations between the different parameters expressing how strong the coupling between the plasma particles is are

$$\gamma = \left(\frac{1}{36 \, \pi \, N_D^2} \right)^{1/3} = \left(\frac{9}{4\pi \, \Lambda^2} \right)^{1/3} = \left(\frac{\Gamma^2}{4 \, \pi} \right)^{1/3}$$

what is shown graphically in the following figure:

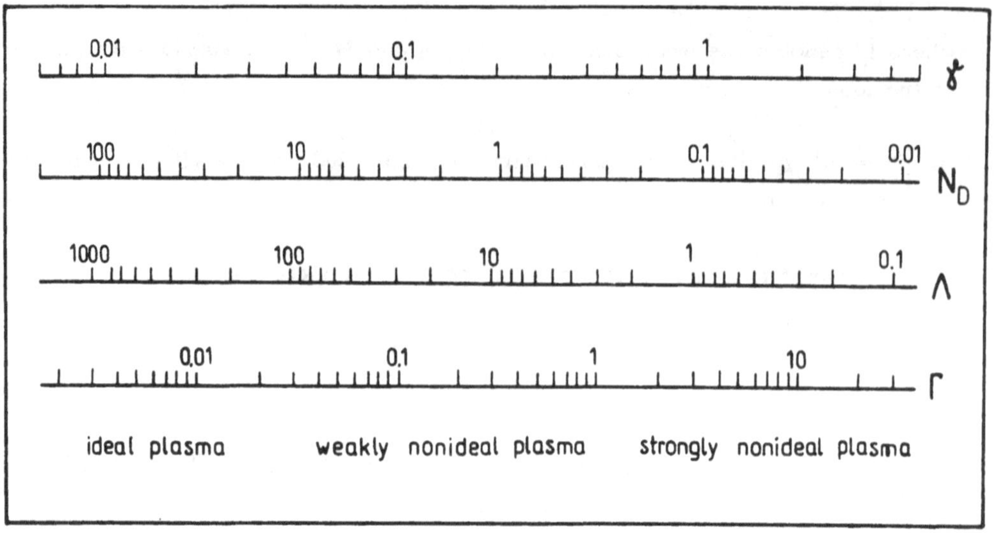

References: [40], [41], [45]

3.5.4. DEGENERATION PARAMETER

The parameter of degeneration, ξ, depends upon the masses of the interacting particles. The electron gas will be degenerate if

$$\xi_e = \frac{E_F}{kT_e} = \frac{(3\pi^2 n_e)^{2/3}\,\hbar^2/2m_e}{kT_e} = \begin{cases} 4.23 \times 10^{-15}\, n_e^{2/3} / T_e^{1/2} & (SI) \\ 4.23 \times 10^{-11}\, n_e^{2/3} / T_e^{1/2} & (CGS) \end{cases} \geq 1$$

And for ions

$$\xi_i = \frac{E_F}{kT_i} = \frac{(3\pi^2 n_i)^{2/3}\,\hbar^2/2m_i}{kT_i} = \begin{cases} 2.31 \times 10^{-18}\, n_i^{2/3} / T_i^{1/2} & (SI) \\ 2.31 \times 10^{-14}\, n_i^{2/3} / T_i^{1/2} & (CGS) \end{cases} \geq 1$$

	n	T
SI	m^{-3}	K
CGS	cm^{-3}	K

$E_F \longrightarrow$ Fermi energy, the numerical values of the previous formula are valid only for hydrogen.

As the density increases, the electron gas is the first to generate whilst the ion gas remains as a classical particle system. At much higher densities, degeneration of the ion gas occurs (see figure at the next page).

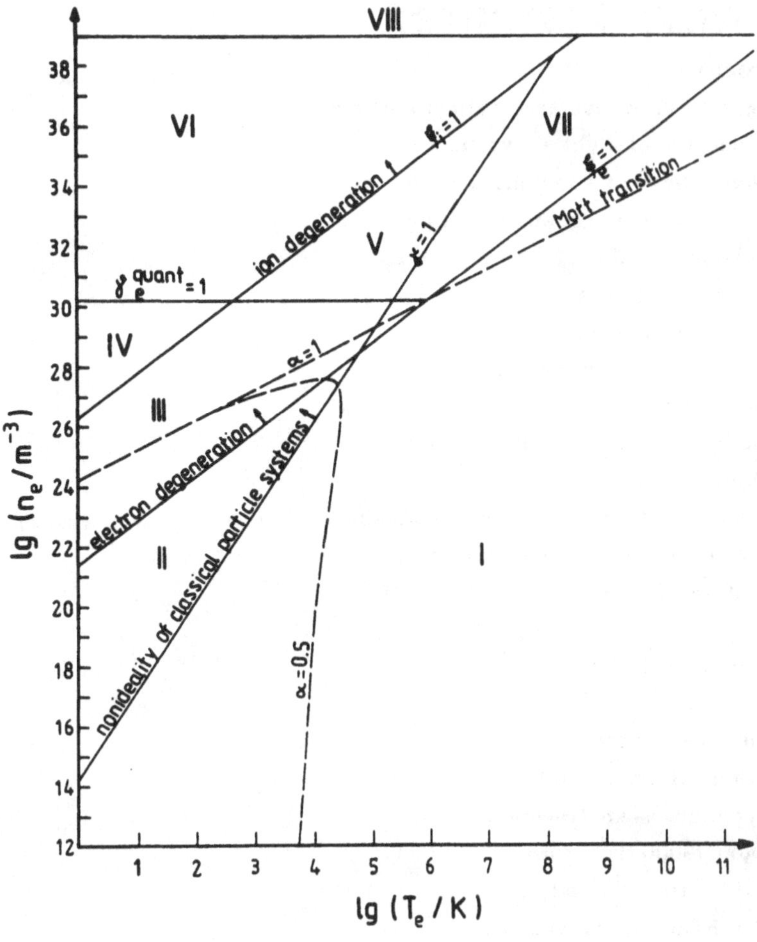

References: [40], [45]

3.5.5. DIMENSIONLESS NUMBERS

Nomenclature

B	magnetic induction (magnetic field strength)
c	→ velocity of light in vacuum
c_p	specific heat at constant pressure
c_v	specific heat at constant volume
$\mathfrak{D}$	diffusion coefficient
e	→ elementary charge
F_g	gravitational force
g	gravitational acceleration
H	specific enthalpy
l	characteristic length of the flow field
m	particle mass
q_{th}	Newton's law heat transfer coefficient. *SI:* $J/m^2 s\ K$, *CGS:* $erg/cm^2 s\ K$
Q	internal energy or reaction energy
v_A	→ Alfvén velocity
v_{fl}	flow velocity
v_s	→ sound velocity
Z	ion charge
η	→ dynamic viscosity
$\varkappa_{th}$	→ thermal conductivity
λ_f	→ particle mean-free-path
λ_{rad}	photon mean-free-path
μ	→ kinematic viscosity
μ_0	→ permeability of vacuum
ν	→ collision frequency
ρ	mass density
σ	→ electrical conductivity
τ	characteristic time of a process
τ_{fl}	characteristic flow time ($= l/v_{fl}$)
ω_c	→ cyclotron frequency

1) Mach number

$$M = \frac{v_{fl}}{v_s} = \frac{\text{flow velocity}}{\text{sound velocity}}$$

The flow is said to be subsonic when $M < 1$

transonic when $M \cong 1$

supersonic when $M > 1$

hypersonic when $M \gg 1$

The Mach number is a measure of gas (plasma) compressibility due to a high flow velocity. Because of the variation of the local Mach number, the flow may be subsonic in one region and supersonic in another part of the flow field.

2) Reynolds number

$$Re = \frac{v_{fl}\, l\, \rho}{\eta} = \frac{v_{fl}\, l}{\mu} = \frac{\text{inertial force}}{\text{viscous force}}$$

When Re is small, the effect of viscosity is important throughout the flow field. When Re is large, the effect of viscosity is important only in a narrow region near a solid boundary or in another transition region. When Re is very large (Re > 10000), the flow becomes turbulent.

3) Prandtl number

$$Pr = \frac{\mu}{(\varkappa_{th}/c_p\, \rho)} = \frac{c_p\, \eta}{\varkappa_{th}} = \frac{\text{kinematic viscosity}}{\text{thermal diffusity}}$$

The smaller μ, the narrower will be the region affected by viscosity. The smaller $\varkappa_{th}/c_p\, \rho$, the narrower will be the region affected by heat conduction. The Prandtl number gives approximately the square of the ratio of the thickness of the velocity boundary layer to that of the thermal boundary layer.

4) Schmidt number

$$Sc = \frac{\mu}{\mathfrak{D}} = \frac{\eta}{\rho\, \mathfrak{D}} = \frac{\text{kinematic viscosity}}{\text{diffusion coefficient}}$$

The Schmidt number is a measure of the diffusion effect. Diffusion phenomena may be characterized by various diffusion coefficients of the various plasma species, or by the corresponding Schmidt numbers of these species.

5) Nusselt number

$$Nu = \frac{q_{th}\, l}{\varkappa_{th}} = \frac{\text{total thermal transfer}}{\text{thermal conduction}}$$

The ratio of the total heat transfer to the conductive heat transfer per unit area and unit time through a surface is referred to as the Nusselt number.

6) Knudsen number

$$Kn = \frac{\lambda_f}{l} = \frac{\text{mean free path}}{\text{characteristic length}}$$

Since the mean-free-path of a gas particle may be taken as $\lambda_f = (\varkappa\ \pi/2)^{1/2}\ \mu/v_s$ the Knudsen number can be re-written as

$$Kn = (\varkappa\ \pi/2)^{1/2}\ \frac{M}{Re}$$

It is convenient to divide the flow field into various regimes:

free molecular flow	$M/Re \geq 3$
transition flow region	$M/Re \leq 3$ & $M/(Re)^{1/2} \geq 1$
slip flow	$0.01 \leq M/(Re)^{1/2} \leq 0.1$
continuum flow	$M/(Re)^{1/2} \leq 0.01$

When the Knudsen number is very small, we have the contimuum flow regime in which the Navier-Stokes equations are a good description of the actual flow. In the slip flow regime, the fluid (plasma) still behaves as a continuous medium but there is a slip of velocity and a jump in temperature at a solid boundary. In the transition region, the flow depends strongly on the discrete character of the plasma. When Kn » 1, the plasma particles should be considered to be free in motion.

7) Knudsen number of radiation

$$Kn_{rad} = \frac{\lambda_{rad}}{l} = \frac{\text{mean free path of a photon}}{\text{charac. length of the flow field}}$$

Kn_{rad} depends on the wavelength (frequency) of radiation;

when $Kn_{rad} \ll 1$ the medium is said to be optically thick

when $Kn_{rad} \gg 1$ the medium is said to be optically thin

8) Peclet number

$$Pe = \frac{v_{fl}\ \ l}{(\varkappa_{th}/c_p\ \rho)} = Pr \circ Re$$

The Peclet number plays the same role in the thermal boundary layer as the Reynolds number in the velocity boundary layer. Whereas Re occurs in momentum transfer terms, Pe occurs in heat transfer terms.

9) Stanton number

$$St = \frac{q_{th}}{v_{fl}\ c_p\ \rho} = \frac{Nu}{Pe} = \frac{\text{total heat transfer}}{\text{convective heat transfer}}$$

10) Froude number

$$Fr = \frac{\rho\ v_{fl}}{F_g\ \ l} = \frac{v_{fl}}{g\ l}$$

Fr is a measure of the gravitational influence on the flow. It is important at a free surface.

11) Hall parameter

$$\omega_c/\nu = \begin{cases} Z\ e\ B\ /\ \nu\ m & (SI) \\ Z\ e\ B\ /\ \nu\ m\ c & (CGS) \end{cases} = \frac{\text{cyclotron frequency}}{\text{collision frequency}}$$

12) Alfven number

$$A = \begin{cases} B/\ v_{fl}\ \ (\mu_0\rho)^{1/2} & (SI) \\ B/\ v_{fl}\ \ (4\pi\ \rho)^{1/2} & (CGS) \end{cases} = \frac{\text{magnetic force}}{\text{inertial force}}$$

13) Lehnert number of the first kind

$$Le_I = Re \circ A = \begin{cases} l\ B\ \rho^{1/2}\ /\ \eta\ \mu_0^{1/2} & (SI) \\ l\ B\ \rho^{1/2}\ /\ \eta\ (4\pi)^{1/2} & (CGS) \end{cases}$$

In magnetohydrodynamics viscosity can be neglected if $Re > 100$ and $Le_I > 100$.

14) Lehnert number of the second kind

$$Le_{II} = Pe \circ A = \begin{cases} l\ B\ c_p\ \rho^{1/2}\ /\ \varkappa_{th}\ \mu_0^{1/2} & (SI) \\ l\ B\ c_p\ \rho^{1/2}\ /\ \varkappa_{th}\ (4\pi)^{1/2} & (CGS) \end{cases}$$

In magnetohydrodynamics thermal conductivity can be neglegted if $Pe > 100$ and $Le_{II} > 100$.

15) Magnetic Reynolds number

$$Re_B = v_{fl}\, l / \mu_B = \begin{cases} v_{fl}\, l\, \sigma\, \mu_0 & (SI) \\ v_{fl}\, l\, \sigma\, 4\pi / c^2 & (CGS) \end{cases} = l/l_B = v_{fl}/v_B$$

Re_B has a similar form to the ordinary Reynolds number, Re, with μ_B replacing the ordinary kinematic viscosity, μ. Re_B describes how the magnetic field will be influenced by the flow motion. When $Re_B \gg 1$, i.e. $l \gg l_B$ and $v_{fl} \ll v_B$, the magnetic field will remain with the flow (frozen-in field). When $Re_B \ll 1$ the magnetic field will not be noticeably influenced by the fluid (plasma) motion.

16) Lundquist number

$$Lu = Re_B \circ A = \begin{cases} \sigma\, l\, B\, (\mu_0/\rho)^{1/2} & (SI) \\ \sigma\, l\, B\, (4\pi/c^2\rho)^{1/2} & (CGS) \end{cases}$$

In magnetohydrodynamics, the electrical conductivity $\sigma \longrightarrow \infty$ holds if $Re_B > 100$ and $Lu > 100$.

17) Magnetic pressure number

$$R_B = \begin{cases} (B^2/2\mu_0)/\ (\rho v_{fl}^2/2) = B^2/(\rho\, v_{fl}^2\, \mu_0) & (SI) \\ (B^2/8\pi)\ /\ (\rho v_{fl}^2/2) = B^2/(\rho\, v_{fl}^2\, 4\pi) & (CGS) \end{cases} = \frac{v_A^2}{v_{fl}^2}$$

$$= \frac{\text{magnetic pressure}}{\text{dynamic pressure}}$$

When $R_B \geq 1$, the plasma will be markedly affected by the magnetic field. If $R_B \ll 1$, the terms due to magnetic field in both the equations of motion and energy may be neglected and the plasma flow will be unaffected by the magnetic field.

18) Hartmann number

$$\text{Ha} = \left(\text{Re} \circ \text{R}_\text{B} \circ \text{Re}_\text{B}\right)^{1/2} = \text{B}\ l\ (\sigma/\eta)^{1/2} = \left(\frac{\sigma\ \text{B}^2\ v_{fl}}{\eta\ v_{fl}/l^2}\right)^{1/2} \qquad (SI)$$

$$= \left(\frac{\text{magnetic force}}{\text{viscous force}}\right)^{1/2}$$

In *CGS*, B should be replaced by $\frac{B}{c}$.

19) Magnetic parameter

$$\text{R}_{mag} = \left(\text{R}_\text{B} \circ \text{Re}_\text{B}\right)^{1/2} = \text{B}\left(\frac{\sigma\ l}{\rho\ v_{fl}}\right)^{1/2} = \left(\frac{\sigma\ \text{B}^2 v_{fl}}{\rho\ v_{fl}^2/l}\right)^{1/2} \qquad (SI)$$

$$= \left(\frac{\text{magnetic force}}{\text{inertial force}}\right)^{1/2}$$

In *CGS*, B should be replaced by $\frac{B}{c}$.

20) Damköhler number of the first kind

$$\mathfrak{D}_I = \tau_{fl}/\tau$$

When $\mathfrak{D}_I \ll 1$, the process with a charcteristic time, τ, is frozen into the flow. In the opposite case, $\mathfrak{D}_I \gg 1$, the flow is in equilibrium at each point in space with respect to the process.

21) Damköhler number of the second kind

$$\mathfrak{D}_{II} = \mathfrak{D}_I\ \frac{Q}{H}$$

$\mathfrak{D}_{II}$ is an energetic parameter characterizing the role of inner degrees of freedom of the flow. The lower $\mathfrak{D}_{II}$ the more a given process of relaxation may be neglegted for that flow.

22) Ratio of specific heats

$$\varkappa = c_p/c_v = \frac{\text{specific heat at constant pressure}}{\text{specific heat at constant volume}}$$

$\varkappa$ is a measure of the internal complexity of the particles of the plasma. From simple gas kinetic theory we have

$$\varkappa = \frac{N + 2}{N}$$

where N is the number of degrees of freedom of a gas particle. For a mono atomic, ideal gas $N = 3$ and $\varkappa = 5/3$.

References: [38], [46], [37] vol.1, [81]

3.6. PRESSURE

The gas kinetic pressure is given by *Dalton's law*

$$p = \sum_{\alpha} n_\alpha \, k \, T_\alpha$$

where the sum is over all species of particles α occuring in the plasma.

The dynamic pressure is defined by the relation

$$p_{dyn} = \rho \, v_{fl}^2 \, / \, 2$$

where ρ is the mass density and v_{fl} denotes the flow velocity

	p	n_α	T_α	ρ	v_{fl}
SI	Pa	m^{-3}	K	kg/m^3	m/s
CGS	$dyne/cm^2$	cm^{-3}	K	g/cm^3	cm/s

The magnetic pressure (normal to the magnetic field lines) is given by

$$p_B = \begin{cases} B^2 \, / \, 2\mu_0 & (SI) \\ B^2 \, / \, 8\pi & (CGS) \end{cases}$$

where B is in Tesla *(SI)* or in Gauss *(CGS)*.

The radiation pressure of black-body radiation is given by

$$p_{rad} = \frac{4}{3} \frac{\sigma_{St/B}}{c} T^4$$

where $\sigma_{St/B}$ is the → Stefan-Boltzmann constant. Occasionally,

$$\frac{4\,\sigma_{St/B}}{c} = a_{rad} = \begin{cases} 7.57 \times 10^{-16}\ \mathrm{J\ m^{-3}\ K^{-4}} & (SI) \\ 7.57 \times 10^{-15}\ \mathrm{erg\ cm^{-3}\ K^{-3}} & (CGS) \end{cases}$$

is also referred to as the Stefan-Boltzmann constant.
As a rule, in a low-temperature plasma, the radiation pressure is negligibly small in comparision with the gas kinetc pressure.
References: [37] vol.1, [46]

3.7. STRESS TENSOR

Notation: p : pressure (scalar)
$\hat{\Pi}$: pressure tensor
$\hat{\tau}$: stress tensor

$$\delta^{ij} = \begin{cases} 0 & \text{if } i \neq j \\ 1 & \text{if } i = j \end{cases}$$

The stress tensor is defined by

$$\tau^{ij} = p\,\delta^{ij} - \Pi^{ij} \tag{1a}$$

or, more generally,

$$\tau^{ij} = a^{kl}_{ij} \frac{\partial v^k}{\partial x^l} \tag{1b}$$

where a^{kl}_{ij} = const. As a consequence of conservation of angular momentum, the tensor $\hat{\tau}$ is symmetric, $\tau^{ij} = \tau^{ji}$. When the viscous properties of the plasma are isotropic, the number of independent coefficients is reduced to two: μ being the coefficient of viscosity and μ' being the second coefficient of viscosity; and $\hat{\tau}$ can be written as

$$\tau^{ij} = \mu \left(\frac{\partial v^i}{\partial x^j} + \frac{\partial v^j}{\partial x^i} \right) + \mu'\,\delta^{ij} \frac{\partial v^k}{\partial x^k} \tag{2}$$

Here the summation convention is used for repeated indices. Ordinarily

$$\mu' + \frac{2}{3}\mu = 0 \tag{3}$$

and therefore

$$\sum \tau^{ii} = (3\mu' + 2\mu)\ \mathit{div}\ v = 0 \qquad (4)$$

The <u>pressure</u> <u>tensor</u> $\hat{\Pi}$ is given by

$$\Pi^{ij} = \rho\ v^i\ v^j \qquad (5)$$

where v^i is the ith component of the flow velocity. The <u>scalar</u> <u>pressure</u> is related to the pressure tensor through,

$$p = \frac{1}{3}\ \mathrm{Tr}\ \hat{\Pi} = \frac{1}{3} \sum \Pi^{ii} \qquad (6)$$

This reduces tothe ordinary hydrodynamic pressure when $\mathit{div}\ v = 0$ (incompressible plasma) or if eq.(3) is fulfilled.

The pressure tensor can be expressed using the stress tensor,

$$\Pi^{ij} = p\ \delta^{ij} - \tau^{ij} = p\ \delta^{ij} - \mu \left(\frac{\partial v^i}{\partial x^j} + \frac{\partial v^j}{\partial x^i} \right) - \mu' \delta^{ij} \frac{\partial v^k}{\partial x^k} \qquad (7)$$

Cap [37] denotes $\hat{\Pi}_*$ as the stress tensor

$$\Pi_*^{ij} = p\ \delta^{ij} - \mu \left(\frac{\partial v^i}{\partial x^j} + \frac{\partial v^j}{\partial x^i} - \frac{2}{3} \delta^{ij} \frac{\partial v^k}{\partial x^k} \right) - \mu' \delta^{ij} \frac{\partial v^k}{\partial x^k} \qquad (8)$$

With $\hat{\Pi}_*$, eq.(6) reduces to the ordinary hydrodynamic pressure if $\mathit{div}\ v = 0$ or $\mu' = 0$ (i.e. inner degrees of freedom do not play an essential role, e.g. for the flow of a monoatomic gas). The non-diagonal elements are related to viscosity (inner friction, dissipation of kinetic energy of the flow).

When the distribution function of particles in the plasma is isotropic and Maxwellian, $\hat{\Pi}$ becomes simply

$$\hat{\Pi} = \begin{pmatrix} p & 0 & 0 \\ 0 & p & 0 \\ 0 & 0 & p \end{pmatrix}$$

<u>References</u>: [37] vol.1 p.152, [38] p.11 and p.178, [60] p.71, [81] p.17

3.8. MAXWELL'S EQUATIONS

E : electric field strength

D : electric displacement

B : magnetic induction

H : magnetic intensity

ρ : charge density

j : current density

	SI	Gaussian *CGS*
Faraday's law	$\text{rot } E = -\frac{\partial B}{\partial t}$	$\text{rot } E = -\frac{1}{c}\frac{\partial B}{\partial t}$
Ampère's law	$\text{rot } H = \frac{\partial D}{\partial t} + j$	$\text{rot } H = \frac{1}{c}\frac{\partial D}{\partial t} + \frac{4\pi}{c} j$
Poisson equation	$\text{div } D = \rho$	$\text{div } D = 4\pi\rho$
absence of mag-tic monopoles	$\text{div } B = 0$	$\text{div } B = 0$
constitutive relations	$D = \varepsilon\, \varepsilon_0\, E$ $B = \mu\, \mu_0\, H$	$D = \varepsilon\, E$ $B = \mu\, H$

In a plasma, $\mu \simeq 1$.

electric polarization	$P = D - \varepsilon_0 E$	$P = \frac{1}{4\pi}(D - E)$
magnetization	$M = \frac{1}{\mu_0} B - H$	$M = \frac{1}{4\pi}(B - H)$

With these definitions, Faraday's law can be rewritten as

	$\text{rot } E = \frac{1}{\varepsilon_0}(\rho - \nabla P)$	$\text{rot } E = 4\pi(\rho - \nabla P)$

where ∇P is the density of the polarization charge, and, if $\partial D/\partial t = 0$, Ampère's can be rewritten as

	$\text{rot } B = \mu_0 (j + \text{rot } M)$	$\text{rot } B = 4\pi(j + \text{rot } M)$

where rot M is the density of the magnetization current.

Lorentz force on charge q	$F = q\,(E + v \times B)$	$F = q\left(E + \frac{v \times B}{c}\right)$
energy density of the electric field	$\varepsilon_0\, E^2 / 2$	$E^2 / 8\pi$
energy density of the magnetic field	$B_0^2 / 2\,\mu_0$	$B^2 / 8\pi$

4. PLASMA MODELS

4.1. MOTION OF CHARGED PARTICLES IN AN ELECTROMAGNETIC FIELD

In this section it will be assumed that the induced electromagnetic field is negligible in comparision with external fields.

Equation of motion of a charged particle (of species α)

$$m_\alpha \frac{d^2r}{dt^2} = q_\alpha \, (E + \dot{r} \times B) \qquad (SI)$$

In *CGS*, B should be replaced by B/c.

Drift velocity of the guiding centre

(i) general

$$v_{F,\alpha} = \frac{1}{q_\alpha} \frac{F \times B}{B^2}$$

(ii) when F is the gravitational force ($= m_\alpha \, g$)

$$v_{g,\alpha} = \frac{m_\alpha}{q_\alpha} \frac{g \times B}{B^2}$$

(iii) when the electric field is homogeneous

$$v_E = \frac{E \times B}{B^2}$$

Note, v_E is independent of the electric charge.

(iv) when the electric field is inhomogeneous

$$v_{E,\alpha} = \left(1 + \frac{1}{4} r_{L,\alpha}^2 \nabla^2\right) \frac{E \times B}{B^2}$$

($r_{L,\alpha}$ is the → Larmor radius)

(v) when the magnetic field is inhomogeneous

∘ gradient drift

$$v_{\nabla B,\alpha} = \frac{1}{2} \frac{q_\alpha}{e} v_\perp r_{L,\alpha} \frac{B \times \nabla B}{B^2}$$

($v_\perp$ is the particle's velocity perpendicular to B)

∘ drift due to curvature of magnetic field lines

$$v_{R_c,\alpha} = \frac{m_\alpha v_\parallel^2}{q_\alpha} \frac{R_c \times B}{R_c^2 B^2}$$

(R_c is the curvature radius)

◦ generalized expression for the drift due to field gradients and curvature of field lines

$$v_{\nabla B, R_c,\alpha} = \frac{m_\alpha}{q_\alpha} \left(v_\parallel^2 + \frac{1}{2} v_\perp^2 \right) \frac{R_c \times B}{R_c^2 B^2}$$

(vi) polarization drift

$$v_{P,\alpha} = \frac{1}{\omega_{c,\alpha} B} \frac{dE}{dt}$$

References: [37] vol.1, [60], [61], [62] chapter 2

4.2. KINETICS

4.2.1. STATISTICAL WEIGHT AND PARTITION FUNCTION

An atomic level with an angular momentum quantum number J has the statistical weight

$$g = 2J + 1$$

In many-electron atomic systems with → LS-coupling (Russell-Saunders coupling), the statistical weight of a multiplet term with the quantum numbers L and S is

$$g = (2S + 1)(2L + 1)$$

where $L = | \sum l |$ is the orbital angular momentum quantum number
$S = | \sum s |$ is the spin quantum number

Reference: [54]

The partition function is defined as

$$\sigma(T) = \sum_{s=1}^{s_{lim}} g_s \exp\left(- \frac{E_s}{kT} \right)$$

where g_s and E_s denote statistical weight and excitation energy of the excited state, s (ground state s = 1, $E_1 = 0$). The summation is over all possible states, s. The limiting value, s_{lim}, is the highest principal quantum number of a bound state ; it can be calculated for hydrogenic states using

$$s_{lim} \leq \left(\frac{Z_{eff}\, \bar{l}_i}{a_0} \right)^{1/2}$$

where

Z_{eff} : effective charge number of the nucleus acting on the electron in the excited state

a_0 : → Bohr radius

$\bar{l}_i$: → mean ion distance

The partition function defined above are sometimes called the "internal" partition function $\sigma^{int}(T)$.

A detailed discussion of the partition function, the theory of the cut-off value, s_{lim}, and tabulated values of $\sigma(T, \Delta E)$ for different atomic species can be found in the book by Drawin and Felenbok ([27] pp. 229-415).

References: [27], [63]

4.2.2. DISTRIBUTION FUNCTIONS

4.2.2.1. BOLTZMANN DISTRIBUTION

is the equilibrium distribution of excited states and is given by

$$\frac{n_u}{n_1} = \frac{g_u}{g_1} \exp\left(-\frac{E_u}{kT}\right)$$

where

n_u : density of particles in an excited (upper) state *u*

n_1 : density of particles in the ground state

E_u : excitation energy of the state *u*

g_u, g_1: statistical weight of the state *u* and the ground state

The number density, n_u, of an excited state, *u*, can be related to the total number density, n, of atoms in a given ionization state by

$$\frac{n_u}{n} = \frac{g_u}{\sigma} \exp\left(-\frac{E_u}{kT}\right)$$

where σ → (internal) partition function.

4.2.2.2. MAXWELLIAN DISTRIBUTION

The number of particles, dN, in the phase space element $dx\, dy\, dz\, dp_x dp_y dp_z$ is given by

$$dN = \frac{N}{V\,(2\pi\; m\; kT)^{3/2}}\; exp\left(-\frac{p_x^2+p_y^2+p_z^2}{2\; m\; kT}\right)\; dx\; dy\; dz\; dp_x dp_y dp_z$$

where N is the total particle number in the volume V, and where $p_\beta = mv_\beta$, β=x,y,z, denotes the momentum in the direction β. With

$$\frac{p^2}{2\;m} = \frac{1}{2\;m}\left(p_x^2+p_y^2+p_z^2\right) = \frac{m}{2}\,v^2 = \frac{m}{2}\left(v_x^2+v_y^2+v_z^2\right) \qquad \text{and} \qquad n = \frac{N}{V}$$

it follows that

$$\frac{dn}{n} = \left(\frac{m}{2\pi\; kT}\right)^{3/2} exp\left\{-\frac{m}{2\;kT}\left(v_x^2+v_y^2+v_z^2\right)\right\} dv_x dv_y dv_z$$

the Maxwellian distribution function is defined by

$$f_{Maxwell}(v_x,v_y,v_z) = \left(\frac{m}{2\pi\; kT}\right)^{3/2} exp\left\{-\frac{m}{2\;kT}\left(v_x^2+v_y^2+v_z^2\right)\right\}$$

$$= \left(\frac{m}{2\pi\; kT}\right)^{3/2} exp\left(-\frac{m\;v^2}{2\;kT}\right)$$

Often, it is not the distribution of velocity components which is required but the distribution of the absolute value of the velocity. Transforming the function dn/n into spherical coordinates, we obtain

$$dn/n = f^*_{Maxwell}(v)\; dv$$

and by integration over all angles,

$$f^*_{Maxwell}(v) = \frac{4\;v^2}{\pi^{1/2}}\left(\frac{m}{2\;kT}\right)^{3/2} exp\left(-\frac{m\;v^2}{2\;kT}\right)$$

Important relations and definitions:

(i) normalization

$$\int_{-\infty}^{\infty} f_M(v)\; d^3v = \int_0^{\infty} f^*_M(v)\; dv = 1$$

(ii) mean velocity

$$\overline{v} = \int_{-\infty}^{\infty} v\, f_M(v)\, d^3v = \int_{0}^{\infty} v\, f_M^*(v)\, dv = \left(\frac{8\ kT}{\pi\ m}\right)^{1/2} = \frac{2\ v_0}{\pi^{1/2}}$$

(iii) most probable velocity

If $\frac{\partial f^*}{\partial v} = 0$ then

$$v_0 = \left(\frac{2\ kT}{m}\right)^{1/2} = \frac{\sqrt{\pi}}{2}\, \overline{v}$$

(iv) mean velocity in a given direction

Because the distribution is isotropic, it is clear that $\overline{v}_x = 0$. But the mean value is finite

$$\overline{|v_x|} = \int_{-\infty}^{\infty} |v_x|\, f_M(v)\, d^3v = \int_{0}^{\infty} |v_x|\, f_M^*(v)\, dv = \left(\frac{2\ kT}{\pi\ m}\right)^{1/2}$$

(v) mean quadratic velocity

$$\overline{v^2} = \int_{-\infty}^{\infty} v^2\, f_M(v)\, d^3v = \int_{0}^{\infty} v^2\, f_M^*(v)\, dv = \frac{3\ kT}{m} = \frac{3}{2}\, v_0^2$$

(vi) kinetic definition of temperature

$$\frac{m}{2}\, \overline{v^2} = \frac{3}{2}\, kT$$

References: [60], [64]

4.2.2.3. FERMI DISTRIBUTION

The particle density in a degenerate ensemble of particles in the element of the velocity space $dv_x dv_y dv_z$ is given by

$$dn\,(v_x,v_y,v_z) = \frac{2\ m^3}{h^3} \frac{dv_x\ dv_y\ dv_z}{1 + exp\left[\frac{m}{2\ kT}\left(v_x^2 + v_y^2 + v_z^2 - v_F^2\right)\right]}$$

where v_F is the Fermi velocity, defined by

$$v_F^2 = \frac{h^2}{m^2}\left(\frac{3\ n}{8\ \pi}\right)^{2/3}$$

As a rule, the particles under consideration are electrons. The With respect to the absolute value of the velocity

$$v = |v| = \left(v_x^2 + v_y^2 + v_z^2\right)^{1/2}$$

the distribution may be written

$$dn(v) = \frac{8\pi\ m^3}{h^3} \frac{v^2\ dv}{1 + exp\left[\frac{m}{2\ kT}\left(v^2 - v_F^2\right)\right]}$$

In terms of the particle energy it reads

$$dn(E) = \frac{4\pi}{h^3}\,(2\ m)^{3/2} \frac{E\ \ dE}{1 + exp\left[\frac{E\ -\ E_F}{kT}\right]}$$

where E_F is the Fermi energy

$$E_F = \frac{h^2}{2\ m}\left(\frac{3\ n}{8\ \pi}\right)^{2/3} = \frac{\hbar^2}{2\ m}\,(3\pi^2\ n)^{2/3}$$

The expression

$$f_{Fermi}(E,T) = \frac{1}{1 + exp\left[\frac{E\ -\ E_F}{kT}\right]}$$

is called the Fermi function or Fermi distribution.

References: [40], [64], [65]

4.2.2.4. LORENTZ DISTRIBUTION

Under the assumptions of the Lorentz gas model (i.e. neglect of collisions between electrons; ions and neutral atoms are at rest [66], [67]), the solution of the → Boltzmann equation for electrons in a weakly ionized plasma gives the Lorentz distribution of electrons

$$f_{Lorentz}(v_e) = A \exp\left\{ - \int_0^{v_e} m_e \, v' \, dv' \Big/ \left(kT_n + \frac{m_n}{3 \, m_e^2} \frac{e^2 \, E^2 \, \xi(v')}{\nu_{en}^2(v')} \right) \right\}$$

where the constant A is determined from

$$\int f_L(v) \, dv = 1$$

$$\xi(v') = \cos^2\theta + \frac{\nu_{en}^2}{\nu_{en}^2 + \omega_{c,e}^2} \sin^2\theta$$

with

$\omega_{c,e}$: → electron cyclotron frequency

ν_{en} : → electron collision frequency with neutral particles

θ : angle between E_0 and B_0

In the case $\nu_{en}(v)$ = const, the Maxwellian distribution is obtained

$$f_L(v) = \left(\frac{m_e}{2\pi \, kT_{eff}} \right)^{3/2} \exp\left(- \frac{v^2 \, m_e}{2 \, kT_{eff}} \right)$$

with an effective temperature

$$T_{eff} = T_n + \frac{m_n \, e^2 \, E_0^2 \, \xi}{2 \, m_e^2 \, \nu_{en}^2 \, k}$$

For $T_n = 0$ and $\omega_{c,e} = 0$ we obtain with $\nu_{en} = v_e/\lambda_0$ the

4.2.2.5. DRYVESTEYN DISTRIBUTION

$$f_{Dryvesteyn}(v_e) = \frac{1}{\pi \, v_D^3 \, \Gamma(3/4)} \exp\left\{ - \left(\frac{v_e}{v_D} \right)^4 \right\}$$

with

$$v_D^2 = \left(\frac{4\ m_i}{3\ m_e}\right)^{1/2} \frac{e\ E_0\ \lambda_0}{m_e}\ , \qquad \lambda_0 = \left(n_n\ \pi\ a^2\right)^{-1}$$

a denotes the radius of a neutral particle.

4.2.3. MOMENTS OF THE DISTRIBUTION FUNCTION $\int M_\alpha\ v^N\ f_\alpha(v)\ dv$

N	M_α	moment	formula
0	1	particle density	$n_\alpha = \int f_\alpha(v)\ dv$
0	$m_\alpha v^2/2$	kinetic energy	$W_\alpha = (m_\alpha/2) \int v^2\ f_\alpha(v)\ dv$
1	m_α	momentum	$I_\alpha = m_\alpha \int v\ f_\alpha(v)\ dv$
1	$m_\alpha v^2/2$	energy flux density	$w_\alpha = (m_\alpha/2) \int v^2 v\ f_\alpha(v)\ dv$
2	m_α	pressure	$p_\alpha = m_\alpha \int v^2\ f_\alpha(v)\ dv$

Reference: [68]

4.2.4. EQUATIONS OF KINETIC THEORY

(1) (general) kinetic equation

$$\frac{df_\alpha}{dt} \equiv \frac{\partial f_\alpha}{\partial t} + v\ \frac{\partial f_\alpha}{\partial r} + \frac{q_\alpha}{m_\alpha}\ (E + v \times B)\ \frac{\partial f_\alpha}{\partial v} = \left(\frac{\delta f_\alpha}{\delta t}\right)_{col} + S_\alpha$$

where $f_\alpha(r,v,t)$: distribution function of species α

q_α, m_α : charge and mass of particles α

$(q_\alpha/m_\alpha)\ E + v \times B$ acceleration of a particle α in an electromagnetic field (in *CGS*-units: B ⟶ B/c)

$\left(\frac{\delta f_\alpha}{\delta t_\alpha}\right)_{col}$: collision operator

S_α : source of particles α

Henceforth, the species-subscript α is dropped and we assume $S = 0$. From the general kinetic equation the following expressions can be obtained

(2) Vlasov equation

$$\frac{df}{dt} = \left(\frac{\delta f}{\delta t}\right)_{col,Vlasov} = 0$$

Although describing a plasma without collisions between the particles of interest, there is an interaction due to the self-consistent electro- magnetic field.

(3) Boltzmann equation

$$\frac{df}{dt} = \left(\frac{\delta f}{\delta t}\right)_{col,Boltzmann}$$

$$= \int \sigma \, |v_2 - v_1| \left\{ f(r_1,v_1^*,t) \, f(r_2,v_2^*,t) - f(r_1,v_1,t) \, f(r_2,v_2,t) \right\} dv_2$$

where σ denotes the collisional cross-section, the indices 1 and 2 refer to the particles involed in the collision and the asterisk $*$ denotes a value befor the collision takes place. The following assumptions are made for the description of collisions

(i) there are only binary collisions

(ii) only short-range interaction forces apply, i.e. the force is zero for $r > r_{col}$

(iii) Boltzmann's hypotheses of molecular chaos: there is no correlation in motion between the colliding species when $r < r_{col}$

(iv) the interaction force between colliding particles depends only on their separation

(v) external forces can be neglected during collisions

(vi) the one-particle distribution function is unchanged during the collision

The r.h.s. of the Boltzmann equation is called the *Boltzmann collision integral.*

(4) Landau equation

(also: kinetic equation with Landau's collision integral, Fokker-Planck equation in Landau form)

$$\frac{df}{dt} = \left(\frac{\delta f}{\delta t}\right)_{col,\,Landau} = 2\pi \frac{q^4}{m^2} \ln \Lambda \frac{\partial}{\partial v_{1j}} \int \left\{ f(v_2) \frac{\partial f(v_1)}{\partial v_{1j}} - f(v_1) \frac{\partial f(v_2)}{\partial v_{2j}} \right\} w_{ij} \, dv_2$$

where the summation convention is used and

$$w_{ij} = \frac{\partial}{\partial g_i}\frac{\partial}{\partial g_j} g, \quad g = |g| = |v_2 - v_1|$$

The same assumptions as for the Boltzmann's collision integral apply but (ii) reads:
(ii) the interaction force can be described by the → Debye-shielded Coulomb potential.

(5) Fokker-Planck equation

$$\frac{df}{dt} = \left(\frac{\delta f}{\delta t}\right)_{\substack{col,\\ Fokker-Planck}} = -\frac{\partial}{\partial v}\left(f\,\frac{\Delta v}{\tau}\right) + \frac{\partial}{\partial u_i\,\partial u_j}\left(\frac{1}{2}\,f\,\frac{\Delta u_i\,\Delta u_j}{\tau}\right) + \ldots$$

where Δv is the change in velocity due to the collsion, and τ denotes the collision time. The same assumptions apply as for the Landau equation.
The explicit form of the r.h.s. is a complex expression and can be found in [37] vol.1

(6)Lenard-Balescu equation

$$\frac{df}{dt} = \left(\frac{\delta f}{\delta t}\right)_{\substack{col, Lenard-\\ Balescu}} = 2\,\frac{q^4}{m^2}\,n\,\frac{\partial}{\partial v_{1j}}\int Q_{ij}\left\{f(v_2)\,\frac{\partial f(v_1)}{\partial v_{1j}} - f(v_1)\,\frac{\partial f(v_2)}{\partial v_{2j}}\right\} dv_2$$

where the tensor Q_{ij} is defined by

$$Q_{ij}(v_1,v_2) = \int \frac{k_i\,k_j\,\delta(kv_1 - kv_2)}{\left|\varepsilon\left(k,\frac{kv_1}{k}\right)\right|^2 k^4}\,dk$$

$\delta(kv_1 - kv_2)$ denotes Dirac's δ-function, n is the mean particle density and $\varepsilon(k,\omega)$ is the dielectric function. The Lenard-Balescu collision integral deals with both binary collisions and collective phenomena (occurrence of $\varepsilon(k,\omega)$). In the static case, $\varepsilon(k,0)$, one obtains the Landau equation.

References: [37] vol.1 chpt.5, [69] p.175, [70] p.311, [71] p.197, [72], [73]

4.3. MOMENT EQUATIONS AND PLASMA DYNAMICS [162]

The general kinetic equation was introduced at the beginning of section 4.2.4. as

$$\frac{df_\alpha}{dt} \equiv \frac{\partial f_\alpha}{\partial t} + v\,\frac{\partial f_\alpha}{\partial r} + \frac{q_\alpha}{m_\alpha}\,(E + v \times B)\,\frac{\partial f_\alpha}{\partial v} = C_\alpha + S_\alpha \tag{1}$$

where C_α denotes the collision operator. The species-subscript α is dropped henceforth.

Total moments of this equation are obtained as follows: Multiply (1) by $Z_k(v)$, where

$$Z_0 = 1\ ; \qquad Z_1 = m\,v\ ; \qquad Z_2 = \frac{1}{2}\,m\,v^2\ ; \qquad Z_3 = \frac{1}{2}\,m\,v^2\,v \tag{2}$$

use tensor notation and the summation rule, to obtain

k = 0 (continuity)

$$\frac{\partial n}{\partial t} + \frac{\partial}{\partial r}\left(n\ u\right) = S^{(0)} \tag{3}$$

k = 1 (momentum balance)

$$m\,\frac{\partial}{\partial t}\left(n\ u_a\right) + \frac{\partial}{\partial x_b}\,M_{ab} - n\,e\left\{E_a + (u \times B)_a\right\} = R_a^{(1)} + S_a^{(1)} \tag{4}$$

k = 2 (energy balance)

$$\frac{1}{2}\,\frac{\partial}{\partial t}\left(\mathcal{T}\!r\ M\right) + \frac{\partial}{\partial x_a}\,Q_a - n\,e\,u_a\,E_a = R^{(2)} + S^{(2)} \tag{5}$$

k = 3 (energy flux)

$$\frac{\partial}{\partial t}\,Q_a + \frac{\partial}{\partial x_b}\,\mathfrak{C}_{ab} - \frac{1}{2}\,\frac{e}{m}\left(\mathcal{T}\!r\ M\right)E_a - \frac{e}{m}\,M_{ab}\,E_b - \frac{e}{m}\,(Q \times B)_a = R^{(3)} + S^{(3)} \tag{6}$$

Here, an averaged value is defined by $\langle A\rangle \equiv \int A\ f\ dv$, and

$n = \langle f\rangle$	density	(7)
$u = \langle v\ f\rangle\ /\ n$	average velocity	(8)
$M_{ab} = m\ \langle v_a\ v_b\rangle$	momentum stress tensor	(9)
$\mathcal{T}\!r\ M = m\ \langle v^2\rangle$	scalar trace of M_{ab}	(10)

$$Q_a = \frac{1}{2} m \langle v^2 v_a \rangle \qquad \text{energy flow} \tag{11}$$

$$\mathfrak{E}_{ab} = \frac{1}{2} m \langle v^2 v_a v_b \rangle \qquad \text{energy flux tensor} \tag{12}$$

$$R^{(k)} = \langle C Z_k \rangle \qquad \text{collisional friction moments} \tag{13}$$

$$S^{(k)} = \langle S Z_k \rangle \qquad \text{source moments} \tag{14}$$

<u>Decomposed</u> <u>moments</u> of the kinetic equation are obtained as follows:
The velocity v may be decomposed into an average part u and a random part w associated with "thermal" motion, i.e.

$$v = u + w \qquad \text{with} \qquad \langle w \rangle = 0 \tag{15}$$

The equations (9) through (12) then become

<u>(A) fluxes and tensors</u>

$$M_{ab} = m\, n\, u_a u_b + P_{ab} \tag{9'}$$

$$\mathcal{Tr}\, M = m\, n\, u^2 + 3\, p \tag{10'}$$

$$Q_a = \frac{1}{2} m\, n\, u^2 u_a + \frac{5}{2} p\, u_a + \Pi_{ab} u_b + q_a \tag{11'}$$

$$\mathfrak{E}_{ab} = \frac{1}{2} m\, n\, u^2 u_a u_b + \frac{1}{2} u^2 P_{ab} + \frac{3}{2} p\, u_a u_b + u_a q_b + u_b q_a$$
$$+ H_{ab} + u_c G_{abc} + u_a u_c P_{bc} + u_b u_c P_{ac} \tag{12'}$$

where

$$P_{ab} = p\, \delta_{ab} + \Pi_{ab} \qquad \text{pressure tensor} \tag{16}$$

$$p = \frac{1}{3} m \langle w^2 \rangle = \frac{1}{3} \mathcal{Tr}\, P \equiv n\, T \qquad \text{scalar pressure, definition of T} \tag{17}$$

$$\Pi_{ab} = m \langle w_a w_b - \frac{1}{3} w^2 \delta_{ab} \rangle \qquad \text{anisotropic shear (viscosity)} \tag{18}$$

$$q_a = \frac{m}{2} \langle w^2 w_a \rangle \qquad \text{heat conduction flux} \tag{19}$$

$$G_{abc} = m \langle w_a w_b w_c \rangle \tag{20}$$

$$H_{ab} = \frac{m}{2} \langle w^2 w_a w_b \rangle \tag{21}$$

If (16) is used to decompose P_{ab} , we have

$$\mathfrak{C}_{ab} = \frac{1}{2} m n u^2 u_a u_b + H_{ab} + u_c G_{abc} + u_a q_b + u_b q_a$$

$$+ \frac{1}{2} p \left(u^2 \delta_{ab} + 7 u_a u_b\right) + \frac{1}{2} u^2 \Pi_{ab} + u_a u_c \Pi_{bc} + u_b u_c \Pi_{ac} \tag{12''}$$

<u>(B) equations</u>

Using lower-order equations to simplify 4-divergence terms such as

$$\frac{\partial}{\partial t}\left(n u_a\right) + \frac{\partial}{\partial x_b}\left(n u_a u_b\right)$$

the decomposed momentum and energy equations read

<u>momentum balance</u>

$$m n \frac{\partial}{\partial t} u_a + m n u_b \frac{\partial}{\partial x_b} u_a + \frac{\partial p}{\partial x_a} + \frac{\partial}{\partial x_b} \Pi_{ab} - e n \left\{E_a + (u \times B)_a\right\}$$

$$= R_a^{(1)} + S_a^{(1)} - m u_a S^{(0)} \tag{4'}$$

<u>energy balance</u>

$$\frac{3}{2}\frac{\partial p}{\partial t} + \left(p \delta_{ab} + \Pi_{ab}\right) \frac{\partial}{\partial x_a} u_b + \frac{3}{2}\frac{\partial}{\partial x_a}\left(p u_a\right) + \frac{\partial}{\partial x_a} q_a$$

$$= R^{(2)} + S^{(2)} - u_a \left(R_a^{(1)} + S_a^{(1)}\right) + \frac{m}{2} u^2 S^{(0)} \tag{5'}$$

<u>energy flux</u>

$$\frac{\partial}{\partial t}\left(Q_a - \frac{5}{2} n \tau u_a\right) + \frac{5}{2} n u_a \frac{\partial \tau}{\partial t} + \frac{\partial}{\partial x_b} \mathfrak{C}_{ab} - \frac{5}{2}\frac{\tau}{m}\frac{\partial}{\partial x_b} M_{ab}$$

$$- \frac{e}{m}\left\{\left(Q - \frac{5}{2} n \tau u\right) \times B\right\}_a - \frac{e}{m} E_b \left\{M_{ab} - \frac{1}{3}\left(\mathcal{T}\!r\, M\right) \delta_{ab}\right\}$$

$$= R_a^{(3)} + S_a^{(3)} - \frac{5}{2}\frac{\tau}{m}\left(R_a^{(1)} + S_a^{(1)}\right) \tag{6'}$$

where

$$n \tau = \frac{1}{3} \mathcal{T}\!r\, M = \frac{1}{3} m u u^2 + n T \tag{22}$$

4.4. ELECTRO-MAGNETO GAS DYNAMICS

Nomenclature:

T_α : kinetic temperature of species α

p_α : (partial) pressure

ρ_α : mass density

u^i_α : ith component of the flow velocity (i = 1,2,3)

E^i : (or D^i) ith component of the electric field

B^i : (or H^i) ith component of the magnetic flux density

n_α : number density of species α

σ_α : mass source per unit volume of species α

τ^{ij}_α : ij-th component of the stress tensor of species α

X^i_α : i-th component of the body force per unit volume of species α

$\sigma_\alpha Z^i_\alpha$: i-th component of the momentum source per unit volume associated with the mass source σ_α

$\mathcal{E}_\alpha$: energy density of species α

$-\frac{\partial}{\partial x^j} Q^j_\alpha$: energy change by heat conduction and radiation

ϵ_α : energy source per unit volume of the species α

The plasma may be considered as a single fluid. Then gross variables (T, p, ρ, u^i) may be used in place of the partial variables (T_α, p_α, ρ_α, u^i_α) of → plasma dynamics.

They are defined as follows

$n = \sum n_\alpha$	(total) number density
$\rho = \sum \rho_\alpha = \sum m_\alpha n_\alpha$	(total) mass density
$p = \sum p_\alpha$	(total) pressure
$T = (1/n) \sum n_\alpha T_\alpha$	temperature
$u^i = (1/\rho) \sum \rho_\alpha u^i_\alpha$	flow velocity
$\rho_e = \sum \rho_{e,\alpha} = \sum e_\alpha n_\alpha$	excess electric charge density

$$j^i = \sum \rho_{e,\alpha} u^i = \sum \rho_{e,\alpha} (u^i_\alpha - u^i) + u^i \sum \rho_{e,\alpha} = I^i + \rho_e u^i \qquad \text{current density}$$

where

I^i : i–th component of the conduction current

$\rho_e u^i$: i–th component of the convection current

There are 16 variables in this problem: T, p, ρ, u^i, E^i, H^i (or B^i), ρ_e, j^i and 16 fundamental equations:

(1) equation of state of the plasma
(2) equation of continuity of the plasma
(3) equation of motion of the plasma (3 in number)
(4) equation of energy
(5) equation of conservation of electric charge
(6) equation of electric current (Ohm's law) (3 in number)
(7) Maxwell's equations (6 in number)

$$p = n\,k\,T = \rho\,R\,T \tag{1}$$

$$\frac{\partial \rho}{\partial t} + \frac{\partial}{\partial x^i}\left(\rho\, u^i\right) = 0 \tag{2}$$

$$\rho \frac{\mathfrak{D} u^i}{\mathfrak{D} t} \equiv \rho \left(\frac{\partial u^i}{\partial t} + u^j \frac{\partial u^i}{\partial x^i}\right) = - \frac{\partial p_{tot}}{\partial x^i} + \frac{\partial \tau^{ij}}{\partial x^j} + F^i_e + F^i_g \tag{3}$$

$$\frac{\partial}{\partial t}\left(\rho\, \mathscr{E}\right) + \frac{\partial}{\partial x^j}\left(\rho\, \mathscr{E}\, u^j\right) = - \frac{\partial}{\partial x^j}\left(u^j\, p_{tot}\right) + \frac{\partial}{\partial x^j} Q^j + \varepsilon_{tot} \tag{4}$$

$$\frac{\partial \rho_e}{\partial t} + \frac{\partial j^i}{\partial x^i} = 0 \tag{5}$$

$$\frac{\partial j^i}{\partial t} + \frac{\partial}{\partial x^j} \sum_\alpha \left(\rho_{e,\alpha}\, u^i_\alpha\, u^j_\alpha - \frac{e_\alpha}{m_\alpha} \tau^{ij}\right) = \sum_\alpha \frac{e_\alpha}{m_\alpha}\left(X^i_\alpha + \sigma_\alpha Z^i_\alpha\right) \tag{6}$$

$$\frac{1}{\mu_0} \operatorname{rot} B = j + \frac{\partial D}{\partial t} \tag{7'}$$

$$\operatorname{rot} E = - \frac{\partial B}{\partial t} \tag{7''}$$

Additional nomenclature:

$R = k/m$	: specific. gas constant
$p_{tot} = p + p_{rad}$	: total pressure
$F_e = \rho_e E + [j \times B]$	: electromagnetic force
F_g	: nonelectric force (e.g. gravitational force)
$\overline{\mathcal{E}}$	: total energy of the plasma per unit mass
Q	: heat flux involving both heat conduction and radiation
$\varepsilon_{tot} = \varepsilon_e + \varepsilon_{chem}$	: total energy source per unit volume
$\varepsilon_e = E \circ j$	: energy source per unit volume due to the electromagnetic field
ε_{chem}	: energy source per unit volume due to chemical reactions
σ	: → electrical conductivity
ω_c	: → cyclotron frequency
ν_{col}	: → collision frequency

The equation for the current density (6) can be transformed into the generalized Ohm's law in the form

$$j = \sigma \left(E + [u \times B] \right) - \frac{\omega_c}{\nu_{col} B} [j \times B]$$

Frequently

$$j^i = \sigma \left(E^i + (u \times B)^i \right)$$

is used. This approach is correct when the Hall current $\frac{\omega_c}{\nu_{col} B} [j \times B]$ is negligibly small, i.e. for dense plasmas and moderate magnetic fields,

$$\frac{\omega_c}{\nu_{col}} \ll 1$$

For the treatment of plasma flow problems, a number of simplifications are possible by neglecting

- radiation pressure and radiation energy (low-temperature plasma)
- non-electric forces
- displacement current $\partial D/\partial t$ (no high-frequency phenomena)
- excess electric charge ρ_e

With the abbreviations

$$\frac{\mathfrak{D}}{\mathfrak{D}t} = \frac{\partial}{\partial t} + u^j \frac{\partial}{\partial x^j} \qquad \text{and} \qquad (\nabla \circ \tau)^i = \frac{\partial \tau^{ij}}{\partial x^j}$$

the equations of electro-magneto gas dynamics can be written as follows

$$p = n\,k\,T \tag{1}$$

$$\frac{\partial \rho}{\partial t} + \nabla\,(\rho u) = 0 \tag{2}$$

$$\rho \frac{\mathfrak{D}u}{\mathfrak{D}t} = -\,\nabla p + [j \times B] + (\nabla \circ \tau) \tag{3}$$

$$\rho \frac{\mathfrak{D}}{\mathfrak{D}t}\left(c_v\,T\right) = -p\,(\nabla\,u) + (\tau\,\nabla)\,u + \nabla\,(\varkappa\,\nabla\,T) + \frac{j^2}{\sigma} \tag{4}$$

$$j = \sigma\,(E + [u \times B]) \tag{6}$$

$$\frac{1}{\mu_0}\,\mathrm{rot}\,B = j \tag{7'}$$

$$\mathrm{rot}\,E = -\,\frac{\partial B}{\partial t} \tag{7''}$$

$$\mathrm{div}\,B = 0 \tag{5a}$$

$$\mathrm{div}\,E = 0 \tag{5b}$$

Here the equation of charge conservation is replaced by the Maxwell equation (5b). Equations (5a) and (5b) are not independent, hence we have a set of 16 differential equations and the 16 unknowns u, p, ρ, T, E, B.

4.5. MAGNETOGASDYNAMICS (MAGNETOHYDRODYNAMICS)

4.5.1. GENERAL

The term "magnetohydrodynamics" (MHD) has often been used loosely. More correctly, it should be used when the plasma compressibilty is negligible.

In magnetogasdynamics, the plasma is regarded as a single fluid. In addition to the simplifications introduced in the last part of section 4.4., we assume that the energy stored in the electric field is negligible in comparision with that in the magnetic field. This is rensonable when the electrical conductivty is very large, i.e. $\sigma \to \infty$, or when $E_0/v_{fl}\, B_0 < 1$.

The unknowns are reduced to the 9 variables u, p, ρ, T and B. The 9 fundamental equations are

(1) equation of state

(2) equation of mass conservation (continuity equation)

(3) equation of motion (conservation of momentum, 3 in number)

(4) equation of energy

(5) equation of the magnetic field (3 in number)

$$p = n\, k\, T = \rho\, R_g\, T \qquad (1)$$

$$\frac{\partial \rho}{\partial t} + \nabla\, (\rho u) = 0 \qquad (2)$$

$$\rho \frac{\mathcal{D} u}{\mathcal{D} t} = \frac{1}{\mu_0}\, (B\, \nabla)\, B - \nabla \left(p + \frac{B^2}{\mu_0} \right) - (\nabla\, \tau) \qquad (3)$$

$$\rho \frac{\mathcal{D}}{\mathcal{D} t}\, h_0 = \frac{\partial p}{\partial t} + \nabla(u\, \tau) + \nabla(\varkappa_{th} \nabla T) + \frac{\nabla \times B}{\mu_0} \left(\mu_B\, [\nabla \times B] - [u \times B] \right) \qquad (4)$$

$$\frac{\partial B}{\partial t} = \nabla \times [u \times B] - \nabla \times \left[\mu_B\, [\nabla \times B] \right] \qquad (5)$$

where $h_0 = c_p T + u^2/2$: stagnation enthalpy

c_p : specific heat at constant pressure

$\mu_B = 1/(\sigma\, \mu_0)$: magnetic viscosity

$\varkappa_{th}$: → thermal conductivity

τ : → stress tensor

4.5.2. SOME IMPORTANT CASES OF MAGNETOGASDYNAMICS

(i) The → magnetic Reynolds number $Re_{mag} \ll 1$. In this case the magnetic flux density B is practically independent of the flow motion. We may assume that B is given and solve the gasdynamic equations (1)-(4) for u, p, ρ, T.

(ii) The → magnetic pressure number $R_B \ll 1$. In this case the gasdynamic equations are also practically independent of B. We may solve the flow problem using the pure gasdynamic equations

$$p = n\,k\,T = \rho\,R\,T \qquad (1)$$

$$\frac{\partial \rho}{\partial t} + \nabla (\rho\, u) = 0 \qquad (2)$$

$$\rho \frac{\mathcal{D}u}{\mathcal{D}t} = - \nabla p + (\nabla\, \tau) \qquad (3)$$

$$\rho \frac{\mathcal{D}h_0}{\mathcal{D}t} = \frac{\partial p}{\partial t} + \nabla\, (u\, \tau) + \nabla \left(x_{th}\, \nabla T \right) \qquad (4)$$

for u, p, ρ, T. Once these variables are known, the equation of the magnetic flux density (5) may be solved.

(iii) Low → Mach number case, incompressible fluid case (magnetohydrodynamics). The equation of state (1) may be replaced by $\rho = \rho_0$ = const.
The unknowns are then u, p, T, B and may be divided into two groups. One group consists of u, p, B and the second all other variables including T. The fundamental equations for u, p, and B are
(1) condition of incompressibility
(2) equation of motion
(3) equation for the magnetic flux density

$$\nabla u = 0 \qquad (1)$$

$$\rho \frac{\mathcal{D}u}{\mathcal{D}t} = \frac{1}{\mu_o} (B \circ \nabla)\, B - \nabla \left(p + \frac{B^2}{2\, \mu_o} \right) + \mu\, \nabla^2 u \qquad (2)$$

$$\frac{\partial B}{\partial t} = \nabla \times [u \times B] + \mu_B\, \nabla^2 B \qquad (3)$$

where the viscosity μ and the magnetic viscosity μ_B are constant.

From the energy equation we have equations for T as follows

a) if $M \rightarrow 0$ but M^2 (R_B/Re_{mag}) and M^2 R_B are not negligibly small

b) if $M \rightarrow 0$ and M^2 (R_B/Re_{mag}) and M^2 R_B are negligibly small (thermal equation of ordinary gasdynamics)

c) incompressible fluid, M^2 (R_B/Re_{mag}) is not small

d) incompressible fluid, M^2 (R_B/Re_{mag}) is small (thermal equation for an incompressible fluid)

$$\rho \frac{\mathfrak{D}}{\mathfrak{D}t}\left(c_p\ T\right) = \nabla\left(\varkappa_{th}\ \nabla T\right) + \frac{\nabla \times B}{\mu_0}\left(\mu_B\ (\nabla \times B) - (u \times B)\right) \tag{4a}$$

$$\rho \frac{\mathfrak{D}}{\mathfrak{D}t}\left(c_p\ T\right) = \nabla\left(\varkappa_{th}\ \nabla T\right) \tag{4b}$$

$$\rho \frac{\mathfrak{D}}{\mathfrak{D}t}\left(c_v\ T\right) = \nabla\left(\varkappa_{th}\ \nabla T\right) + \frac{\mu_B}{\mu_o}\ [\nabla \times B]^2 \tag{4c}$$

$$\rho \frac{\mathfrak{D}}{\mathfrak{D}t}\left(c_v\ T\right) = \nabla\left(\varkappa_{th}\ \nabla T\right) \tag{4d}$$

(iv) Case of ideal magnetogasdynamics (inviscid, non-heat-conducting and perfectly electrically conducting plasma). In this case, $Re = \infty$, $Re_{mag} = \infty$, Pr = finite (i.e. $\sigma = \infty$, $\varkappa_{th} = 0$, $\mu = 0$). The unknowns are u, ρ, the entropy S (or T or p) and B. The 8 differential equations relating the variables are known as the <u>Lundquist equations</u>

(1) continuity equation

(2) equation of motion

(3) equation for the magnetic flux density

(4) energy equation (with the heat conduction, viscous dissipation and Joule's heat terms set to zero)

$$\frac{\partial \rho}{\partial t} + \nabla\ (\rho\ u) = 0 \tag{1}$$

$$\rho \frac{\mathfrak{D}u}{\mathfrak{D}t} = \frac{1}{\mu_0}\ (B\ \nabla)\ B - \nabla\left(p + \frac{B^2}{2\ \mu_0}\right) \tag{2}$$

$$\frac{\partial B}{\partial t} = \nabla \times [u \times B] \tag{3}$$

$$\frac{\mathfrak{D}S}{\mathfrak{D}t} = 0 \tag{4}$$

For the 8 unknowns u, ρ, p, B, the Lundquist equations can be rewritten as

$$\frac{\partial \rho}{\partial t} + \mathit{div}\ (\rho\ u) = 0 \qquad (1)$$

$$\rho \frac{\mathfrak{D}u}{\mathfrak{D}t} = j \times B - \nabla p \qquad \text{(Euler's flow equation)} \qquad (2)$$

$$\frac{\partial B}{\partial t} = \mathit{rot}\ [u \times B] \qquad (3)$$

$$p\ \rho^{-c_p/c_v} = \text{const.} \qquad (4)$$

These equations are also called <u>equations of ideal magnetohydrodynamics</u> (although the density, ρ, is variable). The more general <u>equations of real magnetohydrodynamics</u> account for viscosity, heat conduction and finite electrical conductivity. These equations have been given earlier at the beginning of section 4.5. for the 9 unknowns u, ρ, p, T, and B. They can be rewritten as

$$p = n\ k\ T = \rho\ R_p\ T \qquad (1)$$

$$\frac{\partial \rho}{\partial t} + \mathit{div}\ (\rho\ u) = 0 \qquad (2)$$

$$\rho \frac{\mathfrak{D}u}{\mathfrak{D}t} = j \times B - \nabla p + \mu\ \Delta u + \left(\mu' + \frac{\mu}{3}\right) \mathit{grad}\ \mathit{div}\ u \qquad (3)$$

(Navier-Stokes flow equation)

$$\frac{\mathfrak{D}}{\mathfrak{D}t}\left(\rho\ c_v T + \rho\ u^2/2\right) = \nabla \left(\varkappa_{th}\ \nabla T\right) + j\ E - \nabla\ (u\ \Pi) \qquad (4)$$

$$\frac{\partial B}{\partial t} = \text{rot}\ (u \times B) + \mu_B\ \Delta B \qquad (5)$$

where μ denotes the viscosity, μ' the bulk viscosity, and Π the total stress-tensor. For an inviscid plasma (*div* u = 0) the Navier-Stokes flow equation takes the form

$$\rho \frac{\mathfrak{D}u}{\mathfrak{D}t} = j \times B - \nabla p + \mu\ \Delta u$$

If the Hall current in the differential equation for the current density cannot be neglected {Hall parameter ω_c/ν = O(1) or O(10)} the equation for the magnetic flux density may be written as

$$-\frac{\partial B}{\partial t} = \frac{1}{\mu_0} \mathit{rot} \frac{1}{\sigma} \mathit{rot}\ B - \mathit{rot}\ [u \times B] + \frac{e}{m\ \mu_0\ \nu} \mathit{rot} - [\mathit{rot}\ B \times B]$$

<u>References</u>: [38], [46]

4.6. MODELS DEVELOPED FOR PLASMA SPECTROSCOPY

4.6.1. THE LOCAL THERMODYNAMIC EQUILIBRIUM (LTE) MODEL

The distribution of population densities of electrons is assumed to be determined exclusively by particle collisions but not by radiation processes. Each collision process is accompanied by its inverse at equal rates (principle of detailed balance).

In the LTE model free electrons have a → Maxwellian velocity distribution (section 4.2.2.2.). Excited states are distributed according to the → Boltzmann distribution (see section 4.2.2.1.) while the ionic composition is determined by the → Saha equation (see section 5.2.2.).

Criteria for the → establishment of LTE are thoroughly discussed in section 5.3.

4.6.2. THE STEADY-STATE CORONA MODEL

Instead of each collision process being balanced by its inverse as in the LTE model, the balance in the corona model is between collisional excitation (and ionization) and spontaneous decay (and radiative recombination). The plasma is assumed to be optically thin. Furthermore it is assumed that variations in the temperature or density take place at a rate that is slow in comparison with the relaxation rate of the atomic processes. Unlike the LTE model, quantities of the corona model depend strongly on atomic cross-sectional data. The corona model refers to plasmas of much lower densities than the LTE model.

Equations:

(i) Free electrons are assumed to have a → Maxwellian velocity distribution.

(ii) The ionization-recombination balance may be written as

$$\frac{n_{Z,1}}{n_{Z+1,1}} = \frac{\alpha^{rad}_{Z+1,1}(T_e)}{S^{(e)}_{Z,1}(T_e)}$$

where

$n_{Z,1}$: ground state number density of particles in the charge state Z

$\alpha^{rad}_{Z+1,1}$: → radiative recombination coefficient (for Z+1 fold charged ions in the ground state)

$S^{(e)}_{Z,1}$: → collisional ionzation coefficient (electron impact)

Note that the distribution of ions in the corona model is independent of the electron density.

(iii) The population density of the excited states are determined by the rate of collisional excitation from the ground level balanced by the rate of spontaneous radiative decay:

$$n_e \, n_{Z,1} \, C^{(e)}_{Z,1u}(T_e) = n_{Z,u} \sum_{l<u} A_{Z,ul}$$

where

$C^{(e)}_{Z,1u}$: → rate coefficient of electron impact excitation of an ion in charge state Z from the ground state to the upper state *u*

$A_{Z,ul}$: → transition probability for a spontaneous radiative decay ($u \longrightarrow l$) of an ion in charge state Z (Einstein coeffient)

The criterion of applicability of the corona model can be written as

$$n_e < \frac{\sum_{l<u} A_{ul}}{F^{(e)}_{ul}}$$

i.e. the probability of radiative decay of level *u* is larger than the probability of collisional de-excitation ($F^{(e)}_{ul}$ is the → rate coefficient for electron impact de-excitation, see section 9.2.). If the electron density is higher than determined by this criterion the corona model breaks down.
Using the expressions for hydrogen-like ions, the criterion reads ([149] p.214)

$$n_e < 5.6 \times 10^8 \, (Z+1)^6 \, T_e^{1/2} \, exp\left(\frac{1162 \, (Z+1)^2}{T_e}\right)$$

where n_e is in cm^{-3} and T_e is in Kelvin.

4.6.3. THE TIME-DEPENDENT CORONA MODEL

If the plasma temperature or density has suddenly been changed the plasma relaxes into a new state. The relaxation time, τ, is determined by the slowest processes which influence the population distribution. It can be approximated by

$$\tau = \frac{1}{n_e \left(S^{(e)}_{Z-1} + \alpha^{rad}_Z\right)}$$

where for S and α the steady-state values are used. These values are about 10^{-12} cm^3 s^{-1} for hydrogen-like as well as for not hydrogen-like ions leading to the simple approximate relation

$$\tau \cong \begin{cases} 10^{18}/n_e & (SI) \\ 10^{12}/n_e & (CGS) \end{cases}$$

	τ	n_e
SI	s	m^{-3}
CGS	s	cm^{-3}

If the plasma lifetime is shorter as or about the same as τ, a time-dependent model must be used.
Since the → electron-electron collision time, $\tau_{ee} = \nu_{ee}^{-1}$, is much shorter than τ, a → Maxwellian velocity distribution of free electrons is assumed.

The population density is described by a balance equation of the form

$$\frac{d}{dt} n_{Z,1} = n_{Z-1,1}\, n_e\, S^{(e)}_{Z-1,1} - n_{Z,1}\, n_e\, S^{(e)}_{Z,1} - n_{Z,1}\, n_e\, \alpha^{rad}_{Z,1} + n_{Z+1,1}\, n_e\, \alpha^{rad}_{Z+1,1}$$

while the rate of change of the population of an ion in an excited level u is given by

$$\frac{d}{dt} n_{Z,u} = n_e\, n_{Z,1}\, C^{(e)}_{Z,1u} - n_{Z,u} \sum_{l<u} A_{Z,ul}$$

4.6.4. THE COLLISIONAL-RADIATIVE MODEL

In contrast to the corona model, stepwise collision processes are included (such as transitions between various upper levels and three-body recombination). The collisional-radiative model is based upon the following assumptions:

(i) the plasma is optically thin
(ii) free electrons have a → Maxwellian velocity distribution
(iii) electron impact ionization from any level is partially balanced by three-body recombination into any level
(iv) transitions between any pair of bound levels are induced by electron impact
(v) radiation is emitted by spontaneous decay and radiative recombination

Under these assumptions, the population of a level u is described by the balance equation

$$\frac{d}{dt} n_{Z,u} = - n_{Z,u} \left\{ n_e S^{(e)}_{Z,u} + n_e \sum_{q>u} C^{(e)}_{Z,uq} + n_e \sum_{l<u} F^{(e)}_{Z,ul} + \sum_{l<u} A_{Z,ul} \right\}$$

$$+ n_e \sum_{l<u} C^{(e)}_{Z,lu} n_{Z,l} + n_e \sum_{q>u} F^{(e)}_{Z,qu} n_{Z,q}$$

$$+ \sum_{q>u} n_{Z,q} A_{Z,qu} + n_e n_{Z+1,1} \left\{ n_e \alpha^{coll}_{Z+1,u} + \alpha^{rad}_{Z+1,u} \right\}$$

where

$\alpha^{coll}_{Z+1,u}$ → three-body recombination coefficient (see section 9.4.)

$\alpha^{rad}_{Z+1,u}$ → radiative recombination coefficient (see section 9.4.4.)

(an electron recombines with an Z+1 fold charged ion into the bound level *u* of the newly formed ion in the charge state Z); all other symbols are already defined.

With increasing quantum number the level spacing becomes closer, the probability of collisional transitions increases while the radiative transition probability becomes smaller. Thus a set of Saha equations of the form

$$\frac{n_{Z+1,1}\, n_e}{n_{Z,u}} = \frac{g_{Z+1,1}}{g_{Z,u}}\, 2 \left(\frac{2\pi\, m_e\, kT_e}{h^2} \right)^{3/2} exp\left(- \frac{E_{Z,u}}{kT_e} \right)$$

can be used to calculate the population density of the upper levels (the large summations in the equation for $d/dt(n_{Z,u})$ can be avoided), $g_{Z,u}$ is the → statistical weight of the level *u* of an ion in charge state Z and $E_{Z,u}$ is the ionization energy of the ion in charge state Z in its level *u*.

Reference: [147]

5. THERMODYNAMICS

5.1. THERMODYNAMIC POTENTIALS

(i) Internal energy $U = U(S,V,N_\alpha)$

S entropy

V volume

N_α number of species α

μ_α chemical potential of species α

Infinitesimal changes in U are described by the fundamental equation

$$\boxed{dU = T\,dS - p\,dV + \sum_\alpha \mu_\alpha\,dN_\alpha}$$

and the following relations hold

$$T = \left(\frac{\partial U}{\partial S}\right)_{V,N_\alpha} \qquad p = -\left(\frac{\partial U}{\partial V}\right)_{S,N_\alpha} \qquad \mu_\alpha = \left(\frac{\partial U}{\partial N_\alpha}\right)_{S,V}$$

(ii) Enthalpy $H = H(S,\ p(T,N_\alpha),\ N_\alpha)$

Fundamental equation: $H = U + p\,V$

$$\boxed{dH = T\,dS + V\,dp + \sum_\alpha \mu_\alpha\,dN_\alpha}$$

$d(p\,V)$ is the differential work done by the system.
At constant pressure p,

$$dH = dQ = dU + p\,dV$$

where Q denotes the amount of heat. The following relations hold

$$T = \left(\frac{\partial H}{\partial S}\right)_{p,N_\alpha} \qquad V = \left(\frac{\partial H}{\partial p}\right)_{S,N_\alpha} \qquad \mu_\alpha = \left(\frac{\partial H}{\partial N_\alpha}\right)_{S,p}$$

The specific enthalpy (heat function per unit mass) is defined by

$$H^* = H/(V\,\rho)$$

Under the assumptions
(i) ions are only singly ionized
(ii) the excitation energy stored in the plasma is small

H^* can be approximated by

$$H^* = \frac{5}{2} R_g T (1 + \alpha) + \alpha R_g \theta$$

where R_g is the → specific gas constant, $\theta = (E_i - \Delta E_i)/k$ is the reduced ionization energy expressed in Kelvin, and α denotes the → degree of ionization.

(iii) Helmholtz free energy $F = F(T,V,N_\alpha)$ (force function)

Fundamental equation: $F = U - T S$

$$= p V + \sum_\alpha \mu_\alpha N_\alpha$$

$$dF = dU - T\, dS - S\, dT$$

$$\boxed{dF = -S\, dT - p\, dV + \sum_\alpha \mu_\alpha\, dN_\alpha}$$

The following relations hold

$$S = \left(\frac{\partial F}{\partial T}\right)_{V,N_\alpha} \qquad p = -\left(\frac{\partial F}{\partial V}\right)_{T,N_\alpha} \qquad \mu_\alpha = \left(\frac{\partial F}{\partial N_\alpha}\right)_{T,V}$$

At constant temperature and particle number, it is

$$dF = - p\, dV$$

where $p\, dV$ is the mechanical work done by the system.

(iv) Gibbs free energy $G = G(T,p,N_\alpha)$

Fundamental equation: $G = U - S T + p V$

$$= H - S T$$

$$\boxed{dG = -S\, dT + V\, dp + \sum_\alpha \mu_\alpha\, dN_\alpha}$$

The following relations hold

$$S = -\left(\frac{\partial G}{\partial T}\right)_{p,N_\alpha} \qquad V = \left(\frac{\partial G}{\partial p}\right)_{T,N_\alpha} \qquad \mu_\alpha = \left(\frac{\partial G}{\partial N_\alpha}\right)_{T,p}$$

Reference: [63]

5.2. PLASMA IN THERMODYNAMIC EQUILIBRIUM

5.2.1. SURVEY

A plasma in complete thermodynamic equilibrium (CTE) is described by

- the → radiation laws of Kirchhoff, Planck, Stefan-Boltzmann, and Wien (see section 10.2.)
- the → Maxwellian energy (or velocity) distribution function (see section 4.2.2.2.)
- the → Boltzmann distribution function of excited states (see 4.2.2.1.)
- the → Saha equation (Saha-Eggert equation) describing the ionic composition of the plasma (see section 5.2.2.)

In CTE, all temperatures are equal, i.e.

$$T = T_e = T_i = T_0 = T_{wall} = T_{exitation} = T_{reaction} = T_{radiation}$$

CTE has no practical importance for laboratory plasmas. Often, not only is the radiation in equilibrium with the matter, but the formulae of Maxwell, Boltzmann, and Saha are locally fulfilled. This leads to the concept of local thermodynamic equilibrium (LTE). Validity conditions for LTE are given in section 5.2.3.

5.2.2. PLASMA COMPOSITION [170]

In CTE (or LTE), the variation of the → free energy vanishes, $\delta F = 0$. Considering free electrons and Z-fold charged ions as the species α in the chemical potential

$$\mu_\alpha = \left(\frac{\partial F}{\partial N_\alpha}\right)_{T,V} = \left(\frac{\partial f}{\partial n_\alpha}\right)_{T,V} \quad , \alpha = e, i^{z+} \; (Z = 0, 1, ..., Z_{max})$$

where f is the free energy density (= F/V).

$\delta F = 0$ leads to a set of coupled mass action laws for the reactions $i^{z+1} + e \rightleftarrows i^z$ in terms of the chemical potential,

$$\mu_{z+1} + \mu_e = \mu_z$$

Separating μ_α into an ideal part and an interaction contribution,

$$\mu_\alpha = \mu_\alpha^{ideal} + \mu_\alpha^{interaction}$$

$$\mu_z^{ideal} = kT\, \ln\left(n_z\, \Lambda_{B,z}^3 / \hat{\sigma}_z\right)$$

one gets a system of coupled (non-ideal) <u>Saha</u> <u>equations</u> (also referred to as <u>Saha-Eggert</u> equations):

$$n_z = n_{z+1} \frac{\hat{\sigma}_z}{\hat{\sigma}_{z+1}} \exp\left(\frac{\mu_e^{ideal} - \Delta I_z}{k\,T}\right)$$

where

$$\Lambda_{B,\alpha} = \left(\frac{h^2}{2\pi\, m_\alpha\, kT}\right)^{1/2}$$

is the thermal De Broglie wavelength of species α,

$$\Delta I_z = \mu_z^{interaction} - \mu_{z+1}^{interaction} - \mu_e^{interaction}$$

represents the reduction of the ionization energy due to interactions in the plasma. For the partition functions, the following holds

$$\hat{\sigma}_e = 2 \quad \text{(spin up, spin down)}$$

$$\hat{\sigma}_z = \sum_m g_z^{(m)} \exp\left(-\frac{\hat{E}_z^{(m)}}{kT}\right)$$

where $g_z^{(m)}$ is the statistical weight of particles with charge stage Z in the mth excited state, and $\hat{E}_z^{(m)}$ is the corresponding energy. (Note that the energy is measured with respect to the ground state of the neutral atom, indicated by the symbol ^).

For a non-degenerated electron gas, where

$$\mu_e^{ideal} = kT\, \ln\left(n_e\, \Lambda_{B,e}^3 / \hat{\sigma}_e\right)$$

is valid, one gets the well-known form of the Saha equation

$$\boxed{\frac{n_{z+1}\, n_e}{n_z} = \frac{2\,\sigma_{z+1}}{\sigma_z} \frac{\left(2\pi\, m_e\, kT\right)^{3/2}}{h^3} \exp\left(-\frac{I_z - \Delta I_z}{k\,T}\right)}$$

where I_z is the (ideal) energy required to ionize a particle in charge state Z to charge state Z+1. Here, the partition functions are written without the symbol ^, i.e. they are calculated with respect to the ground state of the corresponding ion.

The numerical form of the Saha equation is

$$\frac{n_{z+1}\, n_e}{n_z} = \frac{\sigma_{z+1}}{\sigma_z} \circ \begin{cases} 4.83 \times 10^{21}\, T^{3/2}\, exp\left((I_z - \Delta I_z)/kT\right) & (SI) \\ \\ 4.83 \times 10^{15}\, T^{3/2}\, exp\left((I_z - \Delta I_z)/kT\right) & (CGS) \end{cases}$$

where

	n	I_z	T
SI	m^{-3}	J	K
CGS	cm^{-3}	erg	K

or

$$\frac{n_{z+1}\, n_e}{n_z} = \frac{\sigma_{z+1}}{\sigma_z} \circ \begin{cases} 6.04 \times 10^{27}\, T^{3/2}\, exp\left((I_z - \Delta I_z)/T\right) & (SI) \\ \\ 6.04 \times 10^{21}\, T^{3/2}\, exp\left((I_z - \Delta I_z)/T\right) & (CGS) \end{cases}$$

where

	n	I_z	T
SI	m^{-3}	eV	eV
CGS	cm^{-3}	eV	eV

If excited states, *m*, of the charge state Z are considered seperately, the Saha equation will have the form

$$\frac{n_{z+1,1}\, n_e}{n_{z,m}} = \frac{2\, g_{z+1,1}}{g_{z,m}} \frac{\left(2\pi\, m_e\, kT\right)^{3/2}}{h^3}\, exp\left(-\frac{I_z^{(m)} - \Delta I_z}{k\, T}\right)$$

where $n_{z+1,1}$ and $g_{z+1,1}$ denote the density and the statistical weight of particles of the charge state Z+1 being in the ground state.

The reduction of the ionization energy in a weakly non-ideal plasma (small and moderate densities, i.e. $n_e < 10^{18}\ cm^{-3}$ at T a few eV) can be described on the basis of Coulomb interaction in the Debye-Hückel model [27], [171]:

$$\Delta I_z = (Z+1)\, K_\varepsilon \frac{e^2}{\lambda_D}$$

with $K_\varepsilon = \begin{cases} 1/(4\pi\, \varepsilon_0) & (SI) \\ 1 & (CGS) \end{cases}$, λ_D → Debye length (see section 3.1.1.).

Numerically this is

$$\Delta I_z = \begin{cases} 1.73 \times 10^{-9}\ (Z+1)/\lambda_D & (SI) \\ 1.73 \times 10^{-7}\ (Z+1)/\lambda_D & (CGS) \end{cases}$$

	ΔI_z	λ_D
SI	eV	m
CGS	eV	cm

or

$$\Delta I_z = 1.39 \times 10^{-3}\ (Z+1)/\lambda_D \qquad (SI) \text{ and } (CGS),$$

	ΔI_z	λ_D
SI	m^{-1}	m
CGS	cm^{-1}	cm

Using the formula for the → Debye length (section 3.1.1.) one obtains

$$\Delta I_z = (Z+1) \left(1 + \frac{\langle Z^2\rangle}{\langle Z\rangle}\right)^{1/2} 2\, \pi^{1/2}\, e^3\, K_\varepsilon^{3/2} \left(\frac{n_e}{kT_e}\right)^{1/2}$$

Using the approximation $\langle Z^2\rangle/\langle Z\rangle \cong Z$, we obtain the following numerical expression

$$\Delta I_z = \begin{cases} 3.34 \times 10^{-30}\ (Z+1)^{3/2} \left[\frac{n_e}{T_e}\right]^{1/2} & (SI) \\ 3.34 \times 10^{-20}\ (Z+1)^{3/2} \left[\frac{n_e}{T_e}\right]^{1/2} & (CGS) \end{cases}$$

	ΔI_z	T_e	n_e
SI	J	K	m^{-3}
CGS	erg	K	cm^{-3}

or

$$\Delta I_z = \begin{cases} 2.09 \times 10^{-11}\ (Z+1)^{3/2} \left[\frac{n_e}{T_e}\right]^{1/2} & (SI) \\ 2.09 \times 10^{-8}\ (Z+1)^{3/2} \left[\frac{n_e}{T_e}\right]^{1/2} & (CGS) \end{cases}$$

	ΔI_z	T_e	n_e
SI	eV	K	m^{-3}
CGS	eV	K	cm^{-3}

or

$$\Delta I_z = \begin{cases} 1.94 \times 10^{-13} \ (Z+1)^{3/2} \left[\dfrac{n_e}{T_e}\right]^{1/2} & (SI) \\ 1.94 \times 10^{-10} \ (Z+1)^{3/2} \left[\dfrac{n_e}{T_e}\right]^{1/2} & (CGS) \end{cases}$$

	ΔI_z	T_e	n_e
SI	eV	eV	m^{-3}
CGS	eV	eV	cm^{-3}

For plasmas at higher densities, when the effects of strongly coupling become marked, quantum mechanical corrections must be included. For a hydrogen plasma, this is easily achieved by introducing the → thermal De Broglie wavelength of electrons (see section 3.1.5.), $\Lambda_{B,e}$, into the formula for ΔI_z, as follows [45],

$$\Delta I_z = (Z+1) \ K_\varepsilon \ \frac{e^2}{(\lambda_D + \Lambda_{B,e}/8)}$$

For strongly coupled plasmas, the <u>Unsöld</u> (or <u>Ecker–Kröll</u>) <u>approximate formula</u> can be used [27], [172], [173]

$$\Delta I_z = (Z+1)^{2/3} \ K_\varepsilon \ 3 \ e^2 \left(\frac{4\,\pi}{3} \ n_e\right)^{1/3}$$

Numerically,

$$\Delta I_z = \begin{cases} 6.964 \times 10^{-9} \ (Z+1)^{2/3} \ n_e^{1/3} & (SI) \\ 6.964 \times 10^{-7} \ (Z+1)^{2/3} \ n_e^{1/3} & (CGS) \end{cases}$$

	ΔI_z	n_e
SI	eV	m^{-3}
CGS	eV	cm^{-3}

or

$$\Delta I_z = \begin{cases} 1.116 \times 10^{-27} \ (Z+1)^{2/3} \ n_e^{1/3} & (SI) \\ 1.116 \times 10^{-18} \ (Z+1)^{2/3} \ n_e^{1/3} & (CGS) \end{cases}$$

	ΔI_z	n_e
SI	J	m^{-3}
CGS	erg	cm^{-3}

or

$$\Delta I_z = 5.617 \times 10^{-3} \ (Z+1)^{2/3} \ n_e^{2/3}$$ *(SI)* and *(CGS)*,

	ΔI_z	n_e
SI	m^{-1}	m^{-3}
CGS	cm^{-1}	cm^{-3}

For the calculation of composition of a LTE plasma we need the set of Saha equations.

Ionization energies are compiled in section 12.2. Statistical weights may be calculated by term configurations (section 12.1.) Tabulated values of I_z, ΔI_z, λ_D, and σ_z may be found in the book by Drawin and Felenbok [27]. Ionization energies are also tabulated in [28], [29], [34], [52], [78] pp.36-40, and [118].

The equation of state for a plasma is given by

$$p = \sum n_\alpha \mu_\alpha - f$$

where the quantities p, n_α, μ_α, and f are already defined in section 5.2.2. If the plasma also contains molecules, an additional set of dissociation-recombination equations (Guldenberg-Waage mass action law) is required.
References: [27], [28], [45], [63], [171]-[173]

5.3. VALIDITY CONDITIONS FOR LOCAL THERMODYNAMIC EQUILIBRIUM (LTE)

5.3.1. PARTIAL LTE IN TIME-INDEPENDENT AND HOMOGENEOUS PLASMAS (PLTE)

Notation:

n_i : population density calculated on the basis of the collisional-radiative model (the index i is the principal quantum number)

n_i^{Saha} : population density calculated on the basis of the → Saha equation

The departure coefficient is defined by

$$b_i = n_i / n_i^{Saha}$$

A level i is said to be in PLTE when b_i is equal to unity to within 10% or better. Collisional excitation and de-excitation rates between closely spaced levels are sufficiently large (so that $0.9 < b_i \leq 1.0$ for all $k > i$) if

$$n_e \geq 10 \frac{\sum_{h<j}^{j-1=i} A_{jh}}{\langle \sigma v_e \rangle_{j \to i}} \tag{1}$$

and

$$n_e \geq 10 \frac{\sum_{h<i} A_{ih}}{\langle \sigma v_e \rangle_{i \to j}} \tag{2}$$

where

A_{jh} is the Einstein coefficient for spontaneous transition $j \to h$,
$\langle \sigma v_e \rangle$ is the $\to$ collisional de-excitation ($j \to i$) and excitation ($i \to j$) rate.

If the level system is sufficiently hydrogen like, condition (2) is stronger than condition (1). Thus, it is sufficient to consider condition (2) only. There are several explicit versions of (2):

* Griem [87] (semiclassical approach)

$$n_e \geq \frac{10}{2\pi^{1/2}} \frac{Z_{eff}^7}{i^{17/2}} \left(\frac{\alpha_F}{a_0}\right) \left(\frac{kT}{E_H(Z_{eff})}\right)^{1/2} = C_1 \frac{Z_{eff}^7}{i^{17/2}} \left(\frac{kT}{E_H(Z_{eff})}\right)^{1/2} \tag{3}$$

with

$$C_1 = \begin{cases} 7.4 \times 10^{24}\ \mathrm{m}^{-3} & (SI) \\ 7.4 \times 10^{18}\ \mathrm{cm}^{-3} & (CGS) \end{cases}$$

$E_H(Z_{eff})$ is the ionization energy of a hydrogen like particle

$$E_H(Z_{eff}) = K_\varepsilon \frac{Z_{eff}^2 e^2}{a_0} = Z_{eff}^2\ 13.60\ \mathrm{eV}$$

where Z_{eff} denotes the effective charge seen by the bound electron, $\alpha_F \to$ fine-structure constant, $a_0 \to$ Bohr radius. The criterion (3) is only applicable to neutral hydrogen ($Z_{eff} = 1$), singly ionized helium ($Z_{eff} = 2$), doubly ionized lithium ($Z_{eff} = 3$) etc. For non-hydrogenic like systems it is only appropriate for higher excited states since most of these levels are quite close to corresponding hydrogenic levels.

* Drawin [88], [89]

$$n_e \geq C_2 Z_{eff}^3 \frac{u_{ij}}{u_1^{3/2}} \frac{\sum_{h<i} A_{ih}}{f_{ij}} \begin{cases} [\Psi_1(u_{ij})]^{-1} & \text{for atoms} \\ [\Psi_2(u_{ij})]^{-1} & \text{for ions} \end{cases} \tag{4}$$

with

$$C_2 = K_\varepsilon^{-1/2} \frac{10 \quad m_e^{1/2}}{8\, \pi^{1/2}\; e\; a_0^{3/2}} = \begin{cases} 1.15 \times 10^{14} \ \mathrm{s/m^3} & (SI) \\ 1.15 \times 10^{8} \ \mathrm{s/cm^3} & (CGS) \end{cases}$$

where

$u_{ij} = |E_i - E_j|/kT$ energy difference in units of kT between two adjacent levels, $j = i + 1$,

$u_1 = E_H(Z_{eff})/kT$ ionization energy of a hydrogen like particle

f_{ij} absorption oscillator strength

In the case of hydrogen like systems, Drawin's criterion (4) becomes Griem's criterion (3) modified by the quantum mechanical correction functions Ψ_1^{-1} and Ψ_2^{-1}

$$n_e \geq C_1 \frac{Z_{eff}^7}{i^{17/2}} \left(\frac{kT}{E_H(Z_{eff})} \right)^{1/2} \begin{cases} [\Psi_1(u_{ij})]^{-1} & \text{for atoms} \\ [\Psi_2(u_{ij})]^{-1} & \text{for ions} \end{cases} \qquad (5)$$

Compared with Griem's formula, eq. (5) leads to about four times larger electron densities at low and medium quantum numbers, and to about ten times smaller values for n_e when i is large ($i > 8$).

References: [87]-[89]

5.3.2. COMPLETE LTE IN TIME-INDEPENDENT AND HOMOGENEOUS PLASMAS (CLTE)

CLTE is established if condition (1) of section 5.3.1. is fulfilled for any two adjacent levels with the ground state included, i.e. (1) has to be fulfilled for the largest energy gap of the level system.

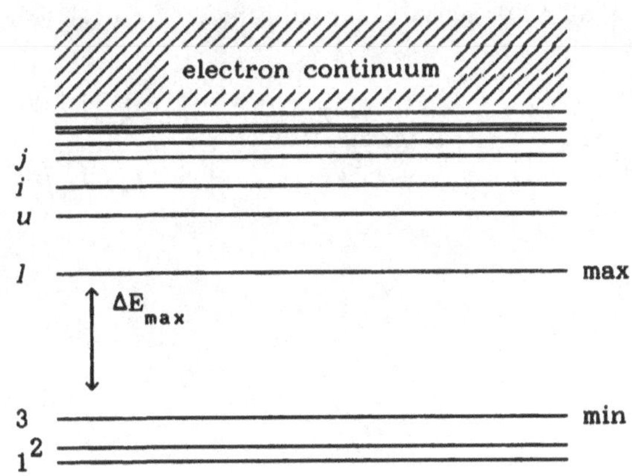

In most cases, the largest energy gap is between the ground state and the first excited state. The indices "min" and "max" refer to the lower and upper levels of the largest energy gap. Then

* Griem [87], for optically thin plasmas

$$n_e \geq \frac{5}{8\,\pi^{1/2}} Z_{eff}^7 \left(\frac{\alpha_F}{a_0}\right)\left(\frac{kT}{E_H(Z_{eff})}\right)^{1/2}\left(\frac{E_{max} - E_{min}}{E_H(Z_{eff})}\right)^3 \tag{6}$$

$$= C_3\, Z_{eff}^7 \left(\frac{kT}{E_H(Z_{eff})}\right)^{1/2}\left(\frac{E_{max} - E_{min}}{E_H(Z_{eff})}\right)^3$$

with

$$C_3 = \begin{cases} 9.25 \times 10^{23}\ \mathrm{m}^{-3} & \textit{(SI)} \\ 9.25 \times 10^{23}\ \mathrm{cm}^{-3} & \textit{(CGS)} \end{cases}$$

* Drawin [88], [89]

$$n_e \geq C_2\, Z_{eff}^3 \frac{g_{max}}{g_{min}} \frac{u_{max}}{u_1^{3/2}} \frac{\sum_{h<max} A_{max\,h}}{f_{min\,max}} \begin{cases} \Phi_1(u_{max}) & \text{for atoms} \\ \Phi_2(u_{min}) & \text{for ions} \end{cases} \tag{7}$$

with $u_{max} = |E_{max} - E_{min}|/kT$, C_2 is already defined in eq.(4), Φ_1 and Φ_2 are quantum mechanical corrections to Griem's formula (6). With respect to a comment by Cillers et al. [90], a more explicit form of eq.(7) may be written

$$n_e \geq C_3\, Z_{eff}^7 \left(\frac{kT}{E_H(Z_{eff})}\right)^{1/2}\left(\frac{E_{max} - E_{min}}{E_H(Z_{eff})}\right)^3 \begin{cases} \Phi_1(u_{max}) & \text{for atoms} \\ \Phi_2(u_{min}) & \text{for ions} \end{cases} \tag{8}$$

References: [87]-[90]

5.3.3. INFLUENCE OF RESONANCE ABSORPTION ON THE ESTABLISHMENT OF PLTE

Even strong resonance absorption does not appreciably change the conditions (3) - (5). The numerical values C_1 and C_2 should be however multiplied by a coefficient not smaller than 0.5.

References: [88], [89]

5.3.4. INFLUENCE OF RESONANCE ABSORPTION ON THE ESTABLISHMENT OF CLTE

As a rule, the first two or three members of the resonance series are completely optically thick in laboratory plasmas. According to Griem [87] and Drawin [88], [89], the r.h.s. of relations (6) - (8) should be reduced by not more than a factor 10, even in case of extremely strong absorption in the resonance lines

$$C_2 \longrightarrow C_2' = C_2/10 \qquad \text{and} \qquad C_3 \longrightarrow C_3' = C_3/10$$

References: [87]-[90]

6. TRANSPORT

6.1. DIFFUSION AND DRIFT

6.1.1. BASIC DESCRIPTION

When the density gradient ∇n_α of species α is small, the diffusion flow J_α is given by Fick's law [109]

$$J_\alpha = - \mathcal{D}_\alpha \nabla n_\alpha$$

where $\mathcal{D}_\alpha$ denotes the scalar diffusion coefficient. Since $J_\alpha = n_\alpha v_\alpha$, Fick's law can be written in the form

$$v_\alpha^{diff} = - \mathcal{D}_\alpha \frac{\nabla n_\alpha}{n_\alpha}$$

The velocity of the center of mass of charged particles α, i.e. the drift velocity v_α^{drift}, is proportional to the electric field intensity E so long as E is weak

$$v_\alpha^{drift} = \mu_\alpha E$$

where μ_α is the scalar mobility (weak-field mobility). In the limit of vanishing electric field and low densities, the mobility μ_α and the diffusion coefficient $\mathcal{D}_\alpha$ are related by the Einstein relation [107] (also referred to as the Nernst-Townsend relation)

$$\frac{\mu_\alpha}{\mathcal{D}_\alpha} = \frac{e}{k\,T}$$

This relation no longer applies in the intermediate-field and high-field region. Here the diffusion coefficient becomes a tensor of the form [111]

$$D = \begin{pmatrix} \mathcal{D}_T & 0 & 0 \\ 0 & \mathcal{D}_T & 0 \\ 0 & 0 & \mathcal{D}_L \end{pmatrix}$$

where $\mathcal{D}_T$ is the transverse and $\mathcal{D}_L$ the longitudinal diffusion coefficient. If, additionally, a magnetic field is present, two of the off-diagonal elements differ from zero [116].

The "low-field" region is defined by

$$\left(\frac{m_\alpha}{m_g} + \frac{m_g}{m_\alpha} \right) e\,E\,\lambda \ll k\,T$$

that is, the energy gained in moving one mean-free-path λ along the field direction is negligible compared with the thermal energy. The factor $(m_\alpha/m_g + m_g/m_\alpha)$ accounts for the inefficiency of energy transfer in elastic collisions when the masses of the particles α and the gas particles are significantly different.

Using the → ideal gas state equation $n\,k\,T = p$, and $\lambda = 1/(n\,\sigma)$, $\sigma \sim 5 \times 10^{-21}\ m^2$, the "low-field" criterion gives

$$\frac{E}{p} \ll 5 \times 10^{-6}\ \frac{\text{statvolt / cm}}{\text{dyne / cm}^2} \cong 2\ \frac{V}{\text{cm Torr}} \cong 1.5\ \frac{V}{\text{m Pa}}$$

Frequently, the fundamental quantity E/n is used instead of E/p. It is common to use

$$1\ \text{Td (Townsend)} = \begin{cases} 10^{-21}\ \text{V m}^2 & \textit{(SI)} \\ 10^{-17}\ \text{V cm}^2 & \textit{(CGS)} \end{cases}$$

as the unit of the reduced electric field E/n [108].

The first approximation of the Chapman-Enskog method, which is valid for vanishing electric field, gives for the diffusion coefficient

$$\mathcal{D}_\alpha = \frac{3}{8}\ \frac{1}{n_g\ \sigma_{\alpha g}}\ \left(\frac{kT}{2\ M}\right)^{1/2}$$

where n_g : number density of gas particles

M : reduced mass of a particle α and a gas particle $\left(= \frac{m_\alpha\ m_g}{m_\alpha + m_g}\right)$

$\sigma_{\alpha g} = \frac{1}{8} \int_0^\infty \exp\left(-\frac{w^2}{2}\right) w^5\ \sigma^*(w)\ dw$, total diffusive cross-section

with

$\sigma^*(w) = \int (1 - \cos\theta)\ d\sigma$, θ the scattering angle

and

$w = \frac{v}{(2\ kT/M)^{1/2}}$ is the normalized relative velocity

In the hard sphere model, $\sigma_{\alpha g} = \pi\ r_g^2$, and ([8] p.36)

$$\mathcal{D}_\alpha = \frac{3}{8}\ \frac{1}{n_g\ r_g^2}\ \left(\frac{kT}{2\pi\ M}\right)^{1/2}$$

If more than the first-order distortion of the distribution function f_α in the field direction has to be considered, the Kihara theory [113] should be applied. Formulaeof this theory and of theories applicable for high-field diffusion such as the two-temperature and the three-temperature theory are complex. They are reviewed e.g. in [110], chapter 5.

References: [32], [107 - 116], [178]

6.1.2. SPREADING OF A CLOUD OF IONS BY DIFFUSION

We consider the spatial dispersion of ions by diffusion through a gas at thermal energy under conditions of low ionization density. The number of ions, S, are located at $t = 0$ at the origin of a one-dimensional coordinate system. The one-dimensional density is given by

$$n(x,t) = \frac{S}{(4\pi \mathcal{D} t)^{1/2}} \exp\left(- \frac{x^2}{4 \mathcal{D} t}\right)$$

which is also referred to as the Einstein relation. From this:

* mean displacement

$$|\bar{x}| = \frac{1}{S} \int_{-\infty}^{\infty} |x| \; n \; dx = \frac{2}{S} \int_{0}^{\infty} x \; n \; dx = \left(\frac{4 \mathcal{D} t}{\pi}\right)^{1/2}$$

* root-mean-square displacement

$$\left(\overline{x^2}\right)^{1/2} = \left(\frac{1}{S} \int_{-\infty}^{\infty} x^2 \; n \; dx\right)^{1/2} = (2 \mathcal{D} t)^{1/2}$$

In three dimensions, the density is given by

$$n(r,t) = \frac{S}{(4\pi \mathcal{D} t)^{1/2}} \exp\left(- \frac{r^2}{4 \mathcal{D} t}\right)$$

* mean displacement

$$\bar{r} = \left(\frac{16 \mathcal{D} t}{\pi}\right)^{1/2}$$

* root-mean-square displacement

$$\left(\overline{r^2}\right)^{1/2} = (6 \mathcal{D} t)^{1/2}$$

The average lifetime τ of ions with respect to collisions with the walls of the containing vessel is

$$\tau = d^2 / \mathcal{D}$$

where d is the characteristic length of the container:

(i) for an infinitely long rectangular tube of width a and depth b

$$\tau = \left[\mathfrak{D}\,\pi^2\left(\frac{1}{a^2} + \frac{1}{b^2}\right)\right]^{-1}$$

(ii) for an infinitely long cylinder of radius R

$$\tau = \frac{1}{\mathfrak{D}}\left(\frac{R}{2.405}\right)^2$$

(iii) for a sphere of radius R

$$\tau = \frac{1}{\mathfrak{D}}\left(\frac{R}{\pi}\right)^2$$

Reference: [110] chapter 1.7

6.1.3. DIFFUSION EQUATION

If there are no scoures or sinks for particles of type α, we obtain from → Fick's law of diffusion and the → equation of continuity

$$\frac{\partial n_\alpha}{\partial t} = \nabla\left(\mathfrak{D}_\alpha\,\nabla n_\alpha\right)$$

This is known as the time-dependent diffusion equation or Fick's second law. If $\mathfrak{D}$ is independent of position, we can separate variables:

$$n_\alpha(x,y,z,t) = n_0(x,y,z)\;T(t)$$

to obtain $T(t) = exp(-t/\tau)$, where τ ia a time constant describing the particle decay by diffusion. The time-independent diffusion equation is given by

$$\nabla^2 n_0 + \frac{n_0}{\mathfrak{D}\,t} = 0$$

where $n_0(x,y,z)$ is the initial steady-state distribution of particles α, established by a continuous supply. The solution for n_0 is an eigenvalue problem depending on the geometry and boundary conditions.

(i) Infinite parallel plates (distance l)

boundary condition:

$$n_0(\pm\, l/2) = 0$$

solution:

$$n(x,t) = \sum A_k \cos\left(\frac{x}{\Lambda_k}\right) \exp\left(-\frac{t}{\tau_k}\right)$$

where the A_k are integration constants, $k = 1,2,3,...,$

$$\Lambda_k = \left(\mathcal{D}\,\tau_k\right)^{1/2} = \frac{1}{2\,k - 1}\,\frac{l}{\pi}$$

are the characteristic length for the kth mode of diffusion. Each mode decay with its own characteristic time constant

$$\tau_k = \tau_1\,(2\,k - 1)^{-2}$$

(ii) **Rectangular parallelepiped** (edge lengths a, b, c)

boundary conditions:

$$n_0 = 0 \quad \text{when} \quad x = \pm\, a/2, \quad y = \pm\, b/2, \quad z = \pm\, c/2$$

solution:

$$n(x,y,z,t) = \sum_{i,j,k=1}^{\infty} G_{ijk} \cos\frac{(2i-1)}{a} \cos\frac{(2j-1)}{b} \cos\frac{(2k-1)}{c} \exp\left(-\frac{t}{\tau_{ijk}}\right)$$

where the G_{ijk} are integration constants.

The diffusion length is $\Lambda_{ijk} = \mathcal{D}\,\tau_{ijk}$, and the time constant

$$\frac{1}{\tau_{ijk}} = \mathcal{D}\,\pi\left[\left(\frac{2i - 1}{a}\right)^2 + \left(\frac{2j - 1}{b}\right)^2 + \left(\frac{2k - 1}{b}\right)^2\right]$$

(iii) **Spherical cavity** (radius R)

boundary condition:

$$n_0(r = 0) < \infty$$

solution:

$$n(r,t) = \sum_{k=1}^{\infty} \frac{B_k}{r} \sin\left(\frac{R}{(\mathcal{D}\,\tau_k)^{1/2}}\right) \exp\left(-\,t/\tau_k\right)$$

where the B_k are integration constants,

$$\frac{R}{(\mathcal{D}\,\tau_k)^{1/2}} = k\,\pi\,, \quad k = 1,2,3,..., \quad \Lambda_k^2 = \mathcal{D}\,\tau_k = (R\,/\,\pi\,k)^2$$

(iv) **Cylindrical cavity (radius R, length l)**

boundary conditions:

$$n_0 = 0 \quad \text{at} \quad r = R \qquad \text{and} \qquad n_0 = 0 \quad \text{at} \quad z = \pm\, l/2$$

solution:

$$n(r,z,t) = \sum_{i,j=1}^{\infty} G_{ij}\, \mathfrak{J}_0(\alpha_i\, r)\, \cos\left(\frac{(2j-1)\,\pi\, z}{l}\right) \exp\left(-\frac{t}{\tau_{ij}}\right)$$

where $\alpha_i R$ is the ith root of $\mathfrak{J}_0$, the Bessel function of the first kind. The diffusion length is given by

$$\frac{1}{\Lambda_{ij}^2} = \frac{1}{\mathcal{D}\,\tau_{ij}} = \alpha_i^2 + \left[\frac{(2j-1)\,\pi}{l}\right]^2$$

Reference: [110] chapter 1

6.1.4. AMBIPOLAR DIFFUSION

For ionization densities of about 10^{14} m^{-3}, Coulomb forces between ions and electrons must be taken into account. Both species of charged particles diffuse with the same velocity [117]

$$v_a = -\,\mathcal{D}_a \frac{\nabla n}{n}$$

where $\mathcal{D}_a$ is the ambipolar diffusion coefficient

$$\mathcal{D}_a = \frac{\mathcal{D}_i\,\mu_e + \mathcal{D}_e\,\mu_i}{\mu_e + \mu_i}$$

$\mathcal{D}_i$, $\mathcal{D}_e$ are the ion and the electron diffusion coefficients, and μ_i, μ_e are the ion and electron mobilities. Using the → Einstein relation and taking $\mu_e \gg \mu_i$ we obtain

$$\mathcal{D}_a = \mathcal{D}_i\left(1 + T_e/T_i\right)$$

and

$$\mathcal{D}_a = \begin{cases} 2\,\mathcal{D}_i = (2\,kT/e)\,\mu_i & \text{for } T_e = T_i = T \\ \mathcal{D}_i\,(T_e/T_i) = (kT_e/e)\,\mu_i & \text{for } T_e \gg T_i \end{cases}$$

The time-dependent diffusion equation for the ambipolar case is

$$\frac{\partial n}{\partial t} = \nabla \left(\mathcal{D}_a \nabla n \right)$$

If $\mathcal{D}_a$ is independent of position and n decays as $\exp(-t/\tau)$, the time independent ambipolar diffusion equation is obtained

$$\nabla^2 n_0 + \frac{n_0}{\mathcal{D}_a \tau} = 0$$

The corresponding diffusion length is given by $\Lambda = (\mathcal{D}_a \tau)^{1/2}$.
References: [32], [110], [112], [117]

6.1.5. DIFFUSUION IN MAGNETIC FIELD

The diffusion perpendicular to the magnetic field lines is described in the classical theory by the classical diffusion coefficient

$$\mathcal{D}_{cl} = \begin{cases} \dfrac{m_e \, kT_e}{e^2 \, B^2} \nu_{ei} = \dfrac{m_e^{1/2} Z^2 e^2 n_i \, \ln \Lambda}{3 \, (2\pi)^{3/2} \, (kT_e)^{1/2} \, B^2} & (SI) \\ \dfrac{m_e \, c^2 \, kT_e}{e^2 \, B^2} \nu_{ei} = \dfrac{4 \, (2\pi)^{1/2} \, m_e^{1/2} Z^2 e^2 n_i \, \ln \Lambda}{3 \, (kT_e)^{1/2} \, B^2} & (CGS) \end{cases}$$

$$= \begin{cases} 1.78 \times 10^{-21} \, Z^2 \, n_i \, \ln \Lambda \, / \, B^2 \, T^{1/2} & (SI) \\ 1.78 \times 10^{-3} \, Z^2 \, n_i \, \ln \Lambda \, / \, B^2 \, T^{1/2} & (CGS) \end{cases}$$

	$\mathcal{D}_{cl}$	Z	n	$\ln \Lambda$	B	T
SI	m^2/s	-	m^3	-	Tesla	K
CGS	cm^2/s	-	cm^3	-	Gauss	K

Experiments have shown that the perpendicular diffusion is better described by the Bohm diffusion coefficient

$$\mathcal{D}_{Bohm} = \begin{cases} \dfrac{1}{16} \dfrac{kT_e}{e \, B} = 5.39 \times 10^{-6} \, T_e/B & (SI) \\ \dfrac{1}{16} \dfrac{c \, kT_e}{e \, B} = 5.39 \times 10^{2} \, T_e/B & (CGS) \end{cases}$$

	$\mathcal{D}_{Bohm}$	T_e	B
SI	m^2/s	K	Tesla
CGS	cm^2/s	K	Gauss

The formulas for $\mathcal{D}_{cl}$ and $\mathcal{D}_{Bohm}$ are only valid for strong fields.

6.2. ELECTRICAL CONDUCTIVITY

The electrical conductivity of a plasma *without* an external magnetic field can be expressed as

$$\sigma = \frac{e^2 n_e}{m_e \sum_h \nu_{eh}}$$

	σ	n_e	ν_{eh}
SI	$(\Omega m)^{-1}$	m^{-3}	s^{-1}
CGS	s^{-1}	cm^{-3}	s^{-1}

where n_e is the electron density and ν_{eh} the → average momentum transfer collision frequency for collisions of electrons with heavy particles of species h.
The electrical conductivity of a completely ionized plasma (Spitzer conductivity) is given by

$$\sigma_{Spitzer} = \gamma_E(Z)\, \sigma_{Lorentz}$$

where $\sigma_{Lorentz}$ is the conductivity of an ideal Lorentz-Plasma

$$\sigma_{Lorentz} = \frac{1}{K_\varepsilon^2} \frac{2\,(2\,k\,T_e)^{3/2}}{\pi^{3/2}\, m^{1/2}\, Z\, e^2\, c^2\, \ln \Lambda}$$

$$= \begin{cases} 2.634 \times 10^{-2}\ \ T_e / Z \ln \Lambda & (SI) \\ 2.634 \times 10^{-13}\ \ T_e / Z \ln \Lambda & (CGS) \end{cases}$$

where T is in Kelvin, and $\gamma_E(Z)$ is a correction factor taking into account e-e interactions. Numerical values are given in the table.

mean ion charge Z	1	2	4	16	∞
$\gamma_E(Z)$	0.582	0.683	0.785	0.923	1.000

References: [31] p.182, [39] p.22, [68] p.22, [74]

The electrical conductivity for a plasma with an external magnetic field is represented by a tensor. In the case of static electric and magnetic fields, neglecting ion motion and for $B_0 \parallel z$ it reads

$$\hat{\sigma} = \begin{pmatrix} \sigma_\perp & \sigma_H & 0 \\ -\sigma_H & \sigma_\perp & 0 \\ 0 & 0 & \sigma_\parallel \end{pmatrix}$$

where

$$\sigma_\perp = \frac{\sigma_\parallel}{1 + \left(\omega_{pl,e}/\nu_e\right)^2}$$

is the transverse conductivity (Pedersen conductivity),

$$\sigma_H = \frac{\omega_{c,e} \quad \sigma_\parallel}{1 + \left[\ 1 + \left(\omega_{c,e}/n_e\right)^2\ \right]}$$

is the Hall conductivity,

$$\sigma_\parallel = \frac{e^2 \ n_e}{m_e \ \nu_e}$$

is the longitudinal conductivity,

$\omega_{pl,e}$ → electron plasma frequency

$\nu_e = \sum \nu_{eh}$ → collision frequency of electrons with heavy particles

Tonks' theorem: The magnetic field B_o does not influence the conductivity parallel to B_o.

In the case of a homogeneous, but time-dependent electric field, $E(T) = E_0 \exp(-i\omega t)$ and a static magnetic field, $B_o \parallel z$, it is

$$\hat{\sigma} = \begin{pmatrix} \sigma_\perp & \sigma_H & 0 \\ -\sigma_H & \sigma_\perp & 0 \\ 0 & 0 & \sigma_\parallel \end{pmatrix}$$

where

$$\sigma_\perp = -i \sum_\alpha \frac{Z_\alpha^2 e^2 n_\alpha (\omega + i\nu_\alpha)}{m_\alpha \left[(\omega + i\nu_\alpha)^2 - \omega_{c,\alpha}^2\right]}$$

$$\sigma_H = -\sum_\alpha \frac{Z_\alpha^2 e^2 n_\alpha \, \omega_{c,\alpha}}{m_\alpha \left[(\omega + i\nu_\alpha)^2 - \omega_{c,\alpha}^2\right]}$$

$$\sigma_\parallel = i \sum_\alpha \frac{Z_\alpha^2 e^2 n_\alpha}{m_\alpha} \frac{1}{(\omega + i\nu_\alpha)}$$

The sum involves all species of charged particles occuring in the plasma. In a two-fluid model α = e, i. If the ion motion can be neglegted, α = e only.

The tensor of the electrical conductivity, $\hat{\sigma}$, is related to the dielectric tensor $\hat{\varepsilon}$ by

$$\hat{\varepsilon} = \mathbf{1} + i\, K_\varepsilon \frac{4\pi}{\omega} \hat{\sigma}$$

where $\mathbf{1}$ denotes the unit tensor, $i = (-1)^{1/2}$, and $K_\varepsilon = \begin{cases} 1/(4\pi\,\varepsilon_0) & (SI) \\ 1 & (CGS) \end{cases}$

This gives

$$\hat{\varepsilon} = \begin{pmatrix} \varepsilon_\perp & \varepsilon_H & 0 \\ -\varepsilon_H & \varepsilon_\perp & 0 \\ 0 & 0 & \varepsilon_\parallel \end{pmatrix}$$

with

$$\varepsilon_\perp = 1 - \sum_\alpha \frac{\omega_{pl,\alpha}^2 (\omega + i\nu_\alpha)}{\omega \left[\left(\omega + i\nu_\alpha\right)^2 - \omega_{c,\alpha}^2\right]}$$

$$\varepsilon_H = -i \sum_\alpha \frac{\omega_{pl,\alpha}^2 \, \omega_{c,\alpha}}{\omega \left[\left(\omega + i\nu_\alpha\right)^2 - \omega_{c,\alpha}^2\right]}$$

$$\varepsilon_\parallel = 1 - \sum_\alpha \frac{\omega_{pl,\alpha}^2}{\omega \left(\omega + i\nu_\alpha\right)}$$

<u>References</u>: [37] vol.1, [75] p.38

6.3. THERMAL CONDUCTIVITY

According to Spitzer [31], the thermal conductivity of a completely ionized plasma can be expressed as

$$\varkappa_{th,Spitzer} = \varepsilon_T(Z)\, \delta_T(Z)\, \varkappa_{th,Lorentz}$$

where

$$\varkappa_{th,Lorentz} = \frac{20}{K_\varepsilon^2} \left(\frac{2}{\pi}\right)^{3/2} \frac{k\ (kT)^{5/2}}{m_e^{1/2}\ Z\ e^4\ \ln\Lambda}$$

$$= \begin{cases} 1.96 \times 10^{-9}\ T^{5/2} / Z \ln\Lambda & (SI) \\ 1.96 \times 10^{-4}\ T^{5/2} / Z \ln\Lambda & (CGS) \end{cases}$$

is the thermal conductivity of a Lorentz plasma, $\varepsilon_T(Z)$ a correction factor considering the secondary electric field due to a temperature gradient and $\delta_T(Z)$ is a correction factor accounting for e-e interactions (see table below).

mean ion charge Z	1	2	4	16	∞
$\delta_T(Z)$	0.225	0.356	0.513	0.791	1.000
$\varepsilon_T(Z)$	0.419	0.410	0.401	0.396	0.400

	$\varkappa_{th}$	Z	$\ln\Lambda$	T
SI	W/K m	-	-	K
CGS	erg/ s K cm	-	-	K

References: [31], [74]

6.4. VISCOSITY

The dynamic viscosity for a completely ionized plasma can be written as

$$\eta = K_\varepsilon^{-2} \frac{0.406\ m_i^{1/2}\ (kT)^{5/2}}{Z^2\ e^4\ \ln\Lambda}$$

$$= \begin{cases} 2.204 \times 10^{-14}\ M_i\ T^{5/2} / (Z^2 \ln\Lambda) & (SI) \\ 2.204 \times 10^{-15}\ M_i\ T^{5/2} / (Z^2 \ln\Lambda) & (CGS) \end{cases}$$

	η	m_i	M_i	T	Z	$\ln \Lambda$
SI	kg/(s m)	kg	-	K	-	-
CGS	g/(s cm)	g	-	K	-	-

where m_i is the ion mass and M_i the corresponding mass number, $\ln \Lambda \rightarrow$ Coulomb logarithm.

Reference: [31]

The kinematic viscosity can be defined by

$$\mu = \lambda_f \left(\frac{2}{\pi \varkappa}\right)^{1/2} v_s$$

	μ	λ_f	v_s
SI	m^2/s	m	m/s
CGS	cm^2/s	cm	cm/s

where $\lambda_f \rightarrow$ mean-free-path of an atom or ion

$\varkappa \rightarrow$ ratio of specific heats

$v_s \rightarrow$ sound velocity

Reference: [46] p.187

The magnetic viscosity (magnetic diffusity) is defined by

$$\mu_B = \begin{cases} 1/(\sigma \mu_o) & (SI) \\ 4\pi/(\sigma c^2). & (CGS) \end{cases}$$

	μ_B	σ
SI	m^2/s	A/Vm
CGS	cm^2/s	s^{-1}

where $\sigma \rightarrow$ electrical conductivity

$\mu_0 \rightarrow$ permeability of vacuum

c $\rightarrow$ velocity of light in vacuum

Reference: [38] p.29

6.5. CLASSICAL TRANSPORT

The transport equations for a multispecies plasma are

$$\frac{\mathcal{D}^\alpha n_\alpha}{\mathcal{D}t} + n_\alpha \nabla\circ v_\alpha = 0$$

$$m_\alpha n_\alpha \frac{\mathcal{D}^\alpha v_\alpha}{\mathcal{D}t} = -\nabla p_\alpha - \nabla\circ \mathsf{P}_\alpha + Z_\alpha e n_\alpha \left\{E + v_\alpha \times B\right\} + R_\alpha$$

$$\frac{3}{2} n_\alpha k \frac{\mathcal{D}^\alpha T_\alpha}{\mathcal{D}t} + p_\alpha \nabla\circ v_\alpha = -\nabla\circ q_\alpha - \mathsf{P}_\alpha : \nabla\circ v_\alpha + Q_\alpha$$

where

$\mathcal{D}^\alpha/\mathcal{D}t = \partial/\partial t + v_\alpha \nabla$

$p_\alpha = n_\alpha kT_\alpha$: partial pressure

$R_\alpha = \sum_\beta R_{\alpha\beta}$: momentum gained by species α through collisions with species β

$Q_\alpha = \sum Q_{\alpha\beta}$: energy gained by species α

P_α : stress tensor

q_α : heat flow

The transport coefficients for a two-component plasma (electrons and singly charged ions) are given below in the limit of large fields ($\omega_{c,\alpha}/\nu_\alpha \gg 1$). The subscripts "$\perp$" and "$\parallel$" refer to the direction of the magnetic field, $B = B\, b$, the subscripts "u" and "th" to the directed and thermal velocity respectively; $u = v_e - v_i$ is the relative streaming velocity, $n \equiv n_e = n_i$ and $j = - n e u$ is the electric current density:

momentum transfer $\quad R_{ei} = -R_{ie} \equiv R = R_u + R_{th}$

frictional force $\quad R_u = n e \left(j_\perp/\sigma_\perp + j_\parallel/\sigma_\parallel\right)$

electrical conductivities $\quad \sigma_\parallel = 2\,\sigma_\perp = 2\,\frac{n e}{m_e \nu_e}$

thermal force $\quad R_{th} = - 0.71\, n k \nabla_\parallel T_e - \frac{3}{2} n k \frac{\nu_e}{\omega_{c,e}} \left\{b \times \nabla_\perp T_e\right\}$

ion heating	$Q_i = 3 \frac{m_e}{m_i} n k \nu_e (T_e - T_i)$
electron heating	$Q_e = - Q_i - R \circ u$
ion thermal flux	$q_i = \varkappa_\parallel^i \nabla_\parallel kT_i - \varkappa_\perp^i \nabla_\perp kT_i + \varkappa_i \left\{ b \times \nabla_\perp kT_i \right\}$
ion thermal conductivities	$\varkappa_\parallel^i = 3.9 \frac{n\ kT_i}{m_i\ \nu_i}, \quad \varkappa_\perp^i = 2 \frac{n\ kT_i\ \nu_i}{m_i\ \omega_{c,i}^2}, \quad \varkappa_i = \frac{5}{2} \frac{n\ kT_i}{m_i\ \omega_{c,i}}$
electron heat flux	$q_e = q_u^e + q_{th}^e$
frictional heat flux	$q_u^e = 0.71\ n\ kT_e\ u_\parallel + \frac{3}{2} n\ kT_e \frac{\nu_e}{\omega_{c,e}} \left\{ b \times u_\perp \right\}$
thermal gradient heat flux	$q_{th}^e = -\varkappa_\parallel^e \nabla_\parallel kT_e - \varkappa_\perp^e \nabla_\perp kT_e - \varkappa_e \left\{ b \times \nabla_\perp kT_e \right\}$
electron thermal conductivities	$\varkappa_\parallel^e = 3.2 \frac{n\ kT_e}{m_e\ \nu_e}, \quad \varkappa_\perp^e = 4.7 \frac{n\ kT_e\ \nu_e}{m_e\ \omega_{c,e}^2}, \quad \varkappa_e = \frac{5}{2} \frac{n\ kT_e}{m_e\ \omega_{c,e}}$
stress tensor (both species, z-axis || B)	$P_{xx} = - \frac{\eta_0}{2} \left(W_{xx} + W_{yy} \right) - \frac{\eta_1}{2} \left(W_{xx} - W_{yy} \right) - \eta_3 W_{xy}$
	$P_{yy} = - \frac{\eta_0}{2} \left(W_{xx} + W_{yy} \right) + \frac{\eta_1}{2} \left(W_{xx} - W_{yy} \right) + \eta_3 W_{xy}$
	$P_{xy} = P_{yx} = - \eta_1 W_{xy} + \frac{\eta_3}{2} \left(W_{xx} - W_{yy} \right)$
	$P_{xy} = P_{yx} = - \eta_2 W_{xz} - \eta_4 W_{yz}$
	$P_{xy} = P_{yx} = - \eta_2 W_{yz} - \eta_4 W_{xz}$
	$P_{xy} = - \eta_0 W_{zz}$

ion viscosity $\quad \eta_0^i = 0.96\ n\ kT_i/\nu_i \quad , \quad \eta_1^i = \frac{3}{10} \frac{n\ kT_i\ \nu_i}{\omega_{c,i}^2}$

$$\eta_2^i = \frac{6}{5} \frac{n\ kT_i\ \nu_i}{\omega_{c,i}^2}, \quad \eta_3^i = \frac{1}{2} \frac{n\ kT_i}{\omega_{c,i}}, \quad \eta_4^i = \frac{n\ kT_i}{\omega_{c,i}}$$

electron viscosity $\quad \eta_0^e = 0.73\ n\ kT_e/\nu_e \quad , \quad \eta_1^e = 0.51 \frac{n\ kT_e\ \nu_e}{\omega_{c,e}^2}$

$$\eta_2^e = 2.0 \frac{n\ kT_e\ \nu_e}{\omega_{c,e}^2}, \quad \eta_3^e = -\frac{1}{2} \frac{n\ kT_e}{\omega_{c,e}}, \quad \eta_4^e = -\frac{n\ kT_e}{\omega_{c,e}}$$

rate-of-strain

tensor (both species) $\quad W_{jk} = \frac{\partial v_j}{\partial x_k} + \frac{\partial v_k}{\partial x_j} - \frac{2}{3} \delta_{jk}\ \nabla \circ v$

* When B = 0 the following simplifications apply:

$$R_u = n\ e\ j/\sigma_\parallel\ , \quad R_{th} = -\ 0.71\ n\ k\ \nabla T_e$$

$$q_i = \chi_\parallel^i\ \nabla\ kT_i, \quad q_u^e = 0.71\ n\ kT_e u\ , \quad q_{th}^e = -\ \chi_\parallel^e\ \nabla\ kT_e$$

$$P_{jk} = -\ \eta_0\ W_{jk}$$

* When $\omega_{c,e}/\nu_e \gg 1 \gg \omega_{c,i}/\nu_i$, the high-field expressions are obeyed by the electrons and the zero-field expressions by the ions.

Collisional transport theory is applicable when

(i) macroscopic time rates of change, $d/dt \ll \nu$; where ν is the smallest collisional frequency,

(ii) * in the absence of a magnetic field: the macroscopic length scales L satisfy $L \gg \lambda$, where $\lambda = \bar{v}/\nu$ is the mean-free-path;

* in a strong field, $\omega_{c,e}/\nu_e \gg 1$, $L_\parallel \gg \lambda$ and $L_\perp \gg (\lambda\ r_{L,e})^{1/2}$ ($L_\perp \gg r_{L,e}$ in a uniform field) where $L_\parallel$ is a macroscopic scale $\parallel$ B and $L_\perp$ is the smaller of $B/|\nabla_\perp B|$ and the transverse plasma dimension.

Additionally, the standard transport coefficients are valid only when

(iii) the → Coulomb logarithm satisfies $\ln \Lambda \gg 1$

(iv) $r_{L,e} \gg \lambda_D$ ($r_{L,e}$ → electron Larmor radius, λ_D → Debye length)
or: $B^2 \ll 8\pi n_e m_e c^2$

(v) relative drift velocity $u = v_\alpha - v_\beta$ is small compared with the thermal velocities, $u^2 \ll kT_\alpha/m_\alpha$

(vi) anomalous transport processes leading to microinstabilities are negligible

Reference: [163]

6.6. TOKAMAK TRANSPORT COEFFICIENTS

6.6.1. BASIC PARAMETERS

The complexity of neoclassical theory prohibits the presentation here of more than a few important results that can be expressed by compact and comprehensible analytic formulae. These are given for the axisymmetric case in the large aspect ratio, circular cross-section approximation. Further refinement may slightly (less than an order of magnitude) modify some of the approximate numerical coefficients.

Basic parameters to be used throughout this chapter are defined as follows (Z_α is the charge state of species α, v_α its thermal velocity; r, R are the minor and the major radius, B_p and B_T are the poloidal and the toroidal magnetic fields of the tokamak respectively).

inverse aspect ratio

$$\varepsilon = r/R \,, \qquad (\varepsilon \ll 1)$$

safety factor

$$q = \varepsilon \, B_T/B_p$$

collisional times (effective 90° Coulomb collisions)

- electron-ion (e-i)

$$\tau_{ei} = (1/Z_i^2) \, \tau_e$$

- ion-impurity (i-I)

$$\tau_{iI} = \frac{n_i}{n_I} \frac{\tau_i}{(Z_i \; Z_I)^2}$$

- self-collision times

$$\tau_{ee} = \sqrt{2}\,\tau_e \,, \qquad \tau_{ii} = Z_i^{-4}\,\tau_i$$

collisionality

$$\nu^*_{\alpha\beta} = \frac{R\,q}{v_\alpha} \frac{1}{\tau_{\alpha\beta}} \,, \qquad \left(\text{where} \quad v_\alpha = (kT_\alpha/m_\alpha)^{1/2}\right)$$

average ("effective") charge

$$\langle Z \rangle = \frac{1}{n_e} \sum_\alpha n_\alpha Z_\alpha^2 \,, \qquad \alpha = i,\, I$$

relative impurity concentration

$$\hat{\alpha} = \left(n_I \; Z_I^2\right) / \left(n_i \; Z_i^2\right)$$

where $\tau_e = \nu_e^{-1}$ and $\tau_i = \nu_i^{-1}$.

6.6.2. NEOCLASSICAL TRANSPORT

In this approximation, the parallel electrical conductivity takes the form

$$\sigma_\parallel^{neo} = \sigma^{cl} \left(1 - \frac{f_T}{1 + \xi\,\nu^*_{ee}/\varepsilon^{3/2}}\right) \left(1 - C_R \frac{f_T}{1 + \xi\,\nu^*_{ee}/\varepsilon^{3/2}}\right)$$

where the number of trapped particles is

$$f_T = 1 - \frac{(1 - \varepsilon)^2}{\left(1-\varepsilon^2\right)^{1/2} \left(1 + 1.46\sqrt{\varepsilon}\right)}$$

with

$$\xi = 0.58 + 0.20\,\langle Z \rangle \,, \qquad C_R = \frac{0.56}{Z_{eff}} \frac{3 - \langle Z \rangle}{3 + \langle Z \rangle}$$

where σ^{cl} is the classical parallel conductivity

$$\sigma^{cl} = \frac{n_e\, e^2\, \tau_e}{m_e} \frac{1}{\langle Z\rangle} \frac{1 + 2.96\ \langle Z\rangle + 0.753\ \langle Z\rangle^2}{1 + 1.19\ \langle Z\rangle + 0.222\ \langle Z\rangle}$$

The following formulae apply to the radial transport only. Several regimes of collisionality may be distinguished:

Pfirsch-Schlüter (PS)	$\nu^* > 1$
plateau (P)	$\varepsilon^{3/2} < \nu^* < 1$
banana (B)	$\nu^* < \varepsilon^{3/2} \quad (\nu^* \ll 1)$
banana-plateau (BP)	$0 < \nu^* < 1$
plateau PS (PPS)	$\nu^* > \varepsilon^{3/2}$

For a two-ion plasma ($m_i \ll m_I$), in the case of weak inter-species temperature equilibration ($\nabla_\parallel(T_i - T_I) \neq 0$) and with the impurity in the PS regime and the (light) ion in any regime (PS or BP), the radial particle and heat fluxes (Γ, q) can be expressed in the following way

$$\Gamma_i = -\frac{Z_I}{Z_i}\Gamma_I = -\mathfrak{D}^{PS}_{iI}\left\{K\left(\frac{\partial n_i}{\partial r} - \frac{Z_i\, n_i}{Z_I\, n_I\, T_i}\frac{\partial p_I}{\partial r}\right) + H\,\frac{n_i}{T_i}\frac{\partial T_i}{\partial r}\right\}$$

$$q_i = -\mathfrak{D}^{PS}_{iI}\,T_i\left\{K_E\left(\frac{\partial n_i}{\partial r} - \frac{Z_i\, n_i}{Z_I\, n_I\, T_i}\frac{\partial p_I}{\partial r}\right) + H_E\,\frac{n_i}{T_i}\frac{\partial T_i}{\partial r}\right\}$$

where $\mathfrak{D}^{PS}_{iI} = r^2_{L,i}\, q^2/\tau_{iI}$ is the PS diffusion coefficient ($\nu^*_{ii} \rightarrow \infty$), and

$$K = 1 - 0.52\,\frac{\hat{\alpha}}{N}$$

$$H = -0.5 + \frac{0.29 + 0.68\,\hat{\alpha}}{N}$$

$$K_E \cong 1$$

$$H_E = 0.5 + \frac{1.41}{\hat{\alpha}} + \frac{0.41\,\hat{\alpha} - 0.34 - 0.17/\hat{\alpha}}{N}$$

$$N = 0.59 + \hat{\alpha} + 1.34\ (2\ \nu^*_{ii})^2$$

$$\hat{\alpha} \ll m_I/m_i$$

In case of $\hat{\alpha} \ll 1$, the relations

$$\mathcal{D}^{PS}_{iI} \cong \mathcal{D}^{PS}_{ii}\ \hat{\alpha}\ H_E \qquad \text{and} \qquad \hat{\alpha}\ H_E = 1.41 - \frac{0.17}{N}$$

are valid. These fluxes must comprise "pure" plasma fluxes due to e-e, e-i, and i-i collisions, according to the different collisionality regimes for electrons and ions (note that in the PS regime, self-collisions are only important for thermal conduction, whilst the Ware-Galeev pinch contribution (~ $E_\parallel$) is only significant in the BP regime). The fluxes are

$$\Gamma_e = Z_i\ \Gamma_i = -\ n_e\ \mathcal{D}^{PS}_e\ \varepsilon^{-3/2} \left(K_{11}\ A_{1e} + K_{12}\ \frac{1}{T_e}\ \frac{\partial T_e}{\partial r}\right) - K_{13}\ \frac{\sqrt{\varepsilon}\ n_e}{B}\ E_\parallel$$

$$q_e = -\ \frac{5}{2}\ T_e\ \Gamma_e - n_e\ \mathcal{D}^{PS}_e\ \varepsilon^{-3/2}\ T_e \left(K_{12}\ A_{1e} + K_{22}\ \frac{1}{T_e}\ \frac{\partial T_e}{\partial r}\right) - K_{13}\frac{\sqrt{\varepsilon}\ n_e\ T_e}{B_p}\ E_\parallel$$

$$q_i = -\ K_4\ T_i\ \Gamma_i - K_2\ n_i\ \mathcal{D}^{PS}_i\ \varepsilon^{-3/2}\ \frac{\partial T_i}{\partial r}$$

where

$$A_{1e} = \frac{1}{p_e}\ \frac{\partial p_e}{\partial r} - \frac{5}{2}\ \frac{1}{T_e}\ \frac{\partial T_e}{\partial r} + \frac{T_i}{Z_i\ T_e} \left(\frac{1}{p_i}\ \frac{\partial p_i}{\partial r} - K_4\ \frac{1}{T_i}\ \frac{\partial T_i}{\partial r}\right)$$

The transport coefficients have the form

$$\mathcal{D}^{PS}_e = r^2_{L,e}\ q^2/\tau_{ei}\ , \qquad \mathcal{D}^{PS}_i = r^2_{L,i}\ q^2/\tau_{ii}$$

$$K_4 = \begin{cases} \dfrac{1.17 - 0.35\left(\nu^*_{ii}/\varepsilon^{3/2}\right)^{1/2}}{\left(1 + 0.7\ (\nu^*_{ii}/\varepsilon^{3/2})^{1/2}\right)\left(1 + (\nu^*_{ee})^2\right)} \equiv M & \text{for } \nu^*_{ii} \ll 1 \\[2ex] \dfrac{M - 2.1\ (\nu^*_{ii})^2}{1 + (\nu^*_{ii})^2} & \text{for } \nu^*_{ii} \gg \varepsilon^{3/2} \end{cases}$$

$$K_2 = 0.66 \left\{ \frac{1}{1 + 1.03 \left(\nu^*_{ii} / \varepsilon^{3/2}\right)^{1/2} + 0.31\, \nu^*_{ii} / \varepsilon^{3/2}} + \frac{1.76\, \nu^*_{ii}\, \varepsilon^{3/2}}{1 + 0.74\, \nu^*_{ii}} \right\}$$

$$K_{mn} = K^{\circ}_{mn} \left\{ \frac{1}{1 + a_{mn} \left(\nu^*_{ei} / \varepsilon^{3/2}\right)^{1/2} + b_{mn}\, \nu^*_{ei} / \varepsilon^{3/2}} + \frac{c^2_{mn}}{b_{mn}} \; \frac{\nu^*_{ii}\, \varepsilon^{3/2}}{1 + c_{mn}\, \nu^*_{ei}} \right\}$$

for m or n = 1,2

$$K_{m3} = K^{\circ}_{m3} \left\{ \frac{1}{\left(1 + a_{m3} \left(\nu^*_{ei} / \varepsilon^{3/2}\right)^{1/2} + b_{m3}\, \nu^*_{ei} / \varepsilon^{3/2}\right) \left(1 + c_{m3}\, \nu^*_{ei}\right)} \right\}$$

The numerical values are summarized in the following table:

Z	m	n	K^0_{mn}	a_{mn}	b_{mn}	c_{mn}
1	1	1	1.04	2.01	1.53	0.89
	1	2	1.20	0.76	0.67	0.56
	2	2	2.55	0.45	0.43	0.43
	1	3	2.30	1.02	1.07	1.07
	2	3	4.19	0.57	0.61	0.61
	3	3	1.83	0.68	0.32	0.66
2	1	1	0.86	2.18	1.17	0.79
	1	2	0.95	0.78	0.50	0.51
	2	2	1.99	0.46	0.26	0.34
	1	3	1.87	0.89	0.62	0.69
	2	3	3.72	0.52	0.34	0.38
	3	3	1.56	0.56	0.25	0.58
4	1	1	0.76	2.30	0.98	0.74
	1	2	0.83	0.80	0.42	0.48
	2	2	1.71	0.46	0.22	0.30
	1	3	1.65	0.79	0.56	0.51
	2	3	3.54	0.48	0.33	0.28
	3	3	1.42	0.47	0.20	0.51

In specific cases the following expressions hold:

$$K_{mn}^{PPS} \cong K_{mn}^{o} \, \mathcal{E}^{3/2} \, \frac{c_{mn}}{b_{mn}} \left\{ 1 + \frac{1}{c_{mn} \, \nu_{ei}^{*} \left(1 - c_{mn} \, \nu_{ei}^{*} \right)} \right\} \qquad \text{for} \quad \nu_{ei}^{*} \gg \mathcal{E}^{3/2}$$

$$K_{2}^{BP} \cong \frac{0.66}{1 + 1.03 \left(\nu_{ii}^{*} / \mathcal{E}^{3/2} \right)^{1/2} + 0.31 \, \nu_{ii}^{*} / \mathcal{E}^{3/2}} \qquad \text{for} \quad \nu_{ii}^{*} \ll 1$$

If the ion thermal flux can be appproximated by $q_i \cong - n_i \, \varkappa_i \, \frac{\partial T_i}{\partial r}$ then, since $\mathcal{D}_e^{PS} \ll \mathcal{D}_i^{PS}$, the ion thermal conductivity reads

$$\varkappa_i = K_2 \, \mathcal{D}_i^{PS} \, \mathcal{E}^{-3/2} = \begin{cases} 0.66 \, \mathcal{D}_i^{PS} \, \mathcal{E}^{-3/2}, & \nu_{ii}^{*} \ll \mathcal{E}^{3/2} \; : \; B \\ 1.6 \, \mathcal{D}_i^{PS} , & \nu_{ii}^{*} \gg 1 \; : \; PS \end{cases}$$

6.6.3. PSEUDOCLASSICAL AND ANOMALOUS TRANSPORT

Considering the toroidal configuration, the pseudoclassical diffusion coefficients can be written in the form

$$\varkappa_e \cong 5 \, \mathcal{D}^{PS} , \qquad \mathcal{D}^{PS} = r_{ep}^{2} \, C / \tau_e , \qquad r_{ep} = \frac{B_T}{B_p} \, r_{L,e}$$

where $r_{L,e}$ is the electron → Larmor radius, C = 1 - 10 eV. They differ from the classical expressions by the factor $\left(B_T / B_p \right)^2 \gg 1$.

The anomalous electron transport is found to scale empirically to the laws given below. Note that $\varkappa_e^{an} \gg \varkappa_e^{neo}$, $\mathcal{D}_e^{an} \cong (0.1 - 0.25) \, \varkappa_e^{an}$.

Alcator $\qquad \varkappa_e = 5 \times 10^{17} / n_e \qquad \mathrm{cm^2 s^{-1}}$

PLT $\qquad \varkappa_e \cong (1/n_e) \, T_e^{\gamma} \quad$ with $\quad (1/2 \le \gamma \le 1)$

T-11 $\qquad \varkappa_e = 10^{20} \, \frac{T_e^{1/2}}{n_e \, q \, R} \, \mathcal{E}^{1.75}$

Ohkawa (theory) $\qquad \varkappa_e = c^2 \, v_{T_e} / (\omega_{pl}^2 \, q \, R) \cong 10^{19} \, \frac{T_e^{1/2}}{n_e \, q \, R} \quad \mathrm{cm^2 s^{-1}}$

7. SHOCKS

7.1. SHOCKS IN IDEAL GAS DYNAMICS

Nomenclature:

D : front velocity in the laboratory frame

v : flow velocity with respect to the shock frame of reference

M : → Mach number

$\varkappa = c_p/c_v$ → ratio of specific heats

$V = 1/\rho$ specific volume

S : entropy

U : specific internal energy

$H = U + p/\rho$ specific enthalpy

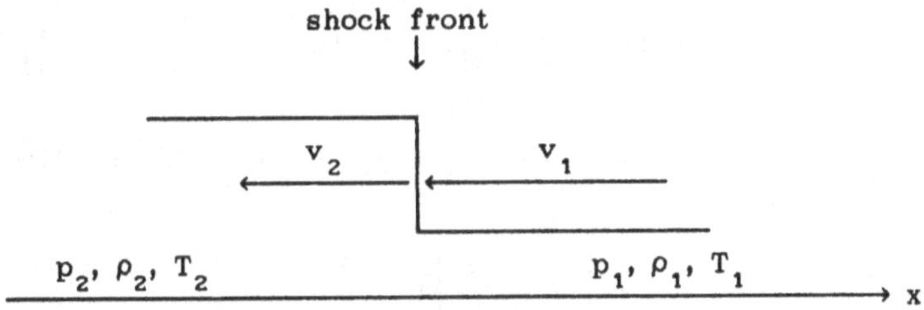

The indices 1 and 2 denote, respectively, quantities in front of and behind the shock front. The subscript n will refer to the component of a vector normal to a shock front surface which is perpendicular to the x-axis. The jump of any physical quantity, Q, across the shock front may be denoted by

$$\{ Q \} = Q_1 - Q_2$$

The relation between the various Q values may be obtained from the differential equations of → gas dynamics or simply by using the conservation laws

mass conservation
$$\left\{ \rho v_n \right\} = 0 \tag{1}$$

conservation of momentum
$$\left\{ v_x^2 \rho + p \right\} = 0$$
$$\left\{ v_x v_y \rho \right\} = 0 \tag{2}$$
$$\left\{ v_x v_z \rho \right\} = 0$$

conservation of energy $$\left\{ \rho\, v_x \left(H + v_x^2/2 \right) \right\} = 0 \qquad (3)$$

From (1) and (2) it follows that

$$v_{x1}^2 = V_1^2 \, \frac{p_2 - p_1}{V_1 - V_2} = \frac{(p_2 - p_1)\, \rho_2}{\rho_1 \rho_2 - \rho_1^2} \qquad (4)$$

($v_1 = -D$ when the gas in front of the shock is not in motion)

$$v_{x2}^2 = V_2^2 \, \frac{p_2 - p_1}{V_1 - V_2} = \frac{(p_2 - p_1)\, \rho_1}{\rho_2^2 - \rho_2 \rho_1} \qquad (5)$$

$$\Delta v_x = v_{x2} - v_{x1} = \left[\left(p_2 - p_1\right) \left(V_1 - V_2\right) \right]^{1/2} = \left[\left(p_2 - p_1\right) \left(\frac{1}{\rho_1} - \frac{1}{\rho_2}\right) \right]^{1/2} \qquad (6)$$

Eliminating v_x, v_y, v_z, eq.(3) gives the <u>Rankine-Hugoniot</u> <u>relation</u> or <u>shock adiabate</u>

$$U_2 - U_1 - \frac{1}{2} \left(p_2 - p_1\right) \left(\frac{1}{\rho_2} + \frac{1}{\rho_1}\right) = 0 \qquad (7)$$

For an ideal gas

$$U = c_v\, T = \frac{1}{\varkappa - 1}\, p\, V \qquad \text{and} \qquad H = c_p\, T = \frac{\varkappa}{\varkappa - 1}\, p\, V$$

so that the shock adiabate may be expressed explicitly by

$$\frac{p_2}{p_1} = \frac{(\varkappa + 1)\, V_1 - (\varkappa - 1)\, V_2}{(\varkappa + 1)\, V_2 - (\varkappa - 1)\, V_1} \qquad (8)$$

or

$$\frac{V_2}{V_1} = \frac{\rho_1}{\rho_2} = \frac{(\varkappa - 1)\, p_2 + (\varkappa + 1)\, p_1}{(\varkappa + 1)\, p_2 + (\varkappa - 1)\, p_1} \qquad (9)$$

Using the equation of state, the ratio of temperatures is

$$\frac{T_1}{T_2} = \frac{p_1\, V_1}{p_2\, V_2} \qquad (10)$$

For very strong shocks ($p_2 \gg p_1$), eqs.(11) - (14) yield

$$\frac{V_1}{V_2} = \frac{\rho_2}{\rho_1} = \frac{\varkappa + 1}{\varkappa - 1} \tag{11}$$

$$\frac{T_1}{T_2} = \frac{\varkappa + 1}{\varkappa - 1} \frac{p_1}{p_2} \tag{12}$$

$$v_1^2 = \frac{\varkappa + 1}{2} p_2 V_1 \tag{13}$$

$$v_2^2 = \frac{(\varkappa - 1)^2}{2 (\varkappa + 1)} p_2 V_1 \tag{14}$$

Using the definition of the → sound velocity for an ideal gas with constant specific heats

$$v_s^2 = \left(\frac{\partial p}{\partial \rho} \right)_S = \varkappa \frac{p}{\rho} = \varkappa \, p \, V$$

the Mach numbers M_1 in front of and M_2 behind the shock may be written as

$$M_1 = \frac{v_1}{v_{s1}} = \left[\frac{(\varkappa - 1) + \left((\varkappa + 1) \, p_2/p_1 \right)}{2 \, \varkappa} \right]^{1/2} \tag{15}$$

$$M_2 = \frac{v_2}{v_{s2}} = \left[\frac{(\varkappa - 1) + \left((\varkappa + 1) \, p_1/p_2 \right)}{2 \, \varkappa} \right]^{1/2} \tag{16}$$

and eqs.(8) - (10) become

$$\frac{p_2}{p_1} = \frac{2 \, \varkappa \, M_1^2 - \varkappa + 1}{\varkappa + 1} \tag{17}$$

$$\frac{\rho_2}{\rho_1} = \frac{V_1}{V_2} = \frac{(\varkappa + 1) \, M_1^2}{(\varkappa - 1) \, M_1^2 + 2} \tag{18}$$

$$\frac{T_2}{T_1} = \frac{\left[2 \, \varkappa \, M_1^2 - \varkappa + 1 \right] \left[(\varkappa - 1) \, M_1^2 + 2 \right]}{(\varkappa + 1)^2 \, M_1^2} \tag{19}$$

Since the entropy of an ideal gas is $S = c_v \, \ell n \left(p \, V^{\varkappa} \right)$, the entropy change across the shock becomes

$$\Delta S = S_2 - S_1 = c_v \, \ell n \left(\frac{p_2}{p_1} \left(\frac{\rho_1}{\rho_2} \right)^{\varkappa} \right)$$

For weak shocks when $\frac{p_2 - p_1}{p_1} \ll 1$

$$\Delta S \rightarrow c_v \frac{2 \varkappa (\varkappa - 1)}{3 (\varkappa + 1)^2} \left(M_1^2 - 1\right)^3$$

Note that even within in the framework of ideal gas dynamics the entropy change is positive, $\Delta S > 0$, i.e. the propagation of shock waves is irreversible.

References: [81], [82], [83]

7.2. SHOCKS IN IDEAL MAGNETOGASDYNAMICS

Using the nomenclature of section 7.1. and introducing B : magnetic flux density, μ_0 → permeability of vacuum, an anologous set of equations can be written down

(1) mass conservation

$$\left\{\rho v_x\right\} = 0$$

(2) conservation of momentum

$$\left\{\rho v_x^2 + p + \frac{B_y^2 + B_z^2 - B_x^2}{2 \mu_0}\right\} = 0$$

$$\left\{\rho v_x v_y - \frac{B_x B_y}{\mu_0}\right\} = 0$$

$$\left\{\rho v_x v_z - \frac{B_x B_z}{\mu_0}\right\} = 0$$

(3) conservation of energy

$$\left\{\rho v_x \left(H + v_x^2/2\right) + v_x \left(\frac{B_y^2 + B_z^2}{\mu_0}\right) - B_x \frac{B_y v_y + B_z v_z}{\mu_0}\right\} = 0$$

(4) conservation of the normal component of the magnetic flux density

$$\left\{B_x\right\} = 0$$

(5) conservation of the tangential components of the electric field

$$\left\{ E_y \right\} = \left\{ v_x B_z - v_z B_x \right\} = 0$$

$$\left\{ E_z \right\} = \left\{ v_x B_y - v_y B_x \right\} = 0$$

The <u>generalized Rankine-Hugoniot relation</u> (<u>shock adiabate of ideal magnetogas-dynamics</u>) may be written as

$$U_2 - U_1 - \frac{1}{2}\left(p_2 - p_1\right)\left(\frac{1}{\rho_2} + \frac{1}{\rho_1}\right) + \frac{1}{4\mu_0}\left[\left(B_{z2} - B_{z1}\right)^2 + \left(B_{y2} - B_{y1}\right)^2\right]\left(\frac{1}{\rho_2} - \frac{1}{\rho_1}\right) = 0$$

<u>References</u>: [37] vol.III p.86, [38] p.127; [84], [85], [86]

7.3. SHOCKS IN RADIATION GAS DYNAMICS

At very high temperatures, when the → radiation pressure is of the order of the → gaskinetic pressure, the motion of the plasma is strongly influenced by radiation. Additional terms occur in the equations of conservation of momentum and energy. For details see: [46], [82].

8. PROPAGATION OF ELECTROMAGNETIC WAVES

8.1. PROPAGATION IN A LORENTZ PLASMA IN THE ABSENCE OF AN EXTERNAL MAGNETIC FIELD

The Lorentz plasma represents a simplified model under the assumptions:

(i) electrons interact by collective space charge forces only

(ii) ions and neutral particles are stationary and form a charge-neutralizing backround

The equation of motion for electrons is given by

$$m_e \frac{\partial v}{\partial t} = - e\, E(t) - \nu\, m_e\, v$$

The force due to the time dependence of the magnetic field may be neglected in comparison with that due to the electric force (ratio = v/c). This is true unless the the plasma approaches relativistic temperatures.

The (complex) Lorentz conductivity can be defined by

$$\sigma_L = \sigma_r + i\,\sigma_i = \frac{n_e\, e^2}{m_e\, (\nu + i\,\omega)} = \frac{n_e\, e^2\, (\nu - i\,\omega)}{m_e\, (\nu^2 + \omega^2)} \xrightarrow[\nu \ll \omega]{} - i\, \frac{n_e\, e^2}{m_e\, \omega}$$

and the (complex) Lorentz dielectric constant as

$$\varepsilon_L = \varepsilon_r - i\,\varepsilon_i = 1 - i\, \frac{\sigma_L}{\varepsilon_0\, \omega} \qquad (SI)$$

$$= 1 - \frac{\omega_{pl}^2}{\omega\, (\omega - i\,\nu)}$$

$$= \left(1 - \frac{\omega_{pl}^2}{\omega^2 + \nu^2}\right) - i \left(\frac{\omega_{pl}^2\, \nu}{\omega\, (\omega^2 + \nu^2)}\right) \xrightarrow[\nu \ll \omega]{} 1 - \frac{\omega_{pl}^2}{\omega^2}$$

In *CGS*, $4\pi\varepsilon_0$ should be replaced by 1, i.e. $\varepsilon_L = 1 - i\, 4\pi\, \sigma_L/\omega$.

Remark: We use here the notation of Heald and Wharton [42]. Other authors may use the sign "±" in their definitions of the imaginary parts of σ_L and ε_L.

Notation:

ω_{pl} : → plasma frequency, in *SI*: $\left(n_e\, e^2 / \varepsilon_0\, m_e\right)^{1/2}$

ω : angular frequency of the propagating wave

ν : → momentum transfer collision frequency

Further definitions are

* the (complex) refractive index

$$\mu_L = \mu_r - i\,\mu_i = \sqrt{\varepsilon_L}$$

with the real refractive index

$$\mu_r = \Re e\left(\sqrt{\varepsilon_L}\right) = \left\{\frac{1}{2}\left(1 - \frac{\omega_{pl}^2}{\omega^2 + \nu^2}\right) + \frac{1}{2}\left[\left(1 - \frac{\omega_{pl}^2}{\omega^2+\nu^2}\right)^2 + \left(\frac{\nu}{\omega}\,\frac{\omega_{pl}^2}{\omega^2+\nu^2}\right)^2\right]^{1/2}\right\}^{1/2}$$

and the attenuation coefficient

$$\mu_i = -\mathcal{I}m\left(\sqrt{\varepsilon_L}\right) = \left\{-\frac{1}{2}\left(1 - \frac{\omega_{pl}^2}{\omega^2 + \nu^2}\right) + \frac{1}{2}\left[\left(1 - \frac{\omega_{pl}^2}{\omega^2+\nu^2}\right)^2 + \left(\frac{\nu}{\omega}\,\frac{\omega_{pl}^2}{\omega^2+\nu^2}\right)^2\right]^{1/2}\right\}^{1/2}$$

* the (complex) propagation coefficient

$$\gamma_L = \alpha + i\,\beta$$

with the attenuation constant

$$\alpha = \frac{\omega}{c}\,\mu_i$$

and the phase constant

$$\beta = \frac{\omega}{c}\,\mu_r$$

The value $\delta = \alpha^{-1}$ is called the attenuation length or skin depth. The phase factor of a wave propagating in the z-direction is $exp(i\omega t - \gamma_L z)$.
Often, $k = -i\gamma_L$ is used as the complex angular wave number, and the phase factor is then given by $exp\left[i(kr - \omega t)\right]$.

* phase velocity

$$v_\Phi = \frac{\omega}{\beta} = \frac{\omega}{k}$$

* group velocity

$$v_g = \frac{d\omega}{d\beta} = \frac{d\omega}{dk}$$

In case of a low-loss plasma ($\nu \ll \omega_{pl}$), three frequency regions are usually distinguished:

(1) low frequencies ($\omega < \nu$)

The conductivity is largely real and approximately

$$\sigma_r \cong \frac{n_e\, e^2}{m_e\, \nu}$$

with

$$\mu_r \cong \left(\frac{\omega_{pl}^2}{2\,\omega\,\nu}\right)^{1/2} \left(1 - \frac{\omega}{2\,\nu}\right)$$

$$\mu_i \cong \left(\frac{\omega_{pl}^2}{2\,\omega\,\nu}\right)^{1/2} \left(1 + \frac{\omega}{2\,\nu}\right)$$

$$\delta \cong \frac{c}{\omega_{pl}} \left(\frac{2\,\nu}{\omega}\right)^{1/2} \left(1 - \frac{\omega}{2\,\nu}\right)$$

(2) intermediate frequencies ($\nu < \omega < \omega_{pl}$)

There is still no propagation of electromagnetic waves in the plasma. Expansion in the limit $\nu \ll \omega \ll \omega_{pl}$ gives

$$\mu_r \cong \frac{\nu\,\omega_{pl}}{2\,\omega^2} \left(1 - \frac{5\,\nu^2}{8\,\omega^2} + \frac{\omega^2}{2\,\omega_{pl}^2}\right)$$

$$\mu_i \cong \frac{\omega_{pl}}{\omega} \left(1 - \frac{3\,\nu^2}{8\,\omega^2} - \frac{\omega^2}{2\,\omega_{pl}^2}\right)$$

$$\delta \cong \frac{c}{\omega_{pl}} \left(1 + \frac{3\,\nu^2}{8\,\omega^2} + \frac{\omega^2}{2\,\omega_{pl}^2}\right)$$

(3) high frequencies ($\nu \ll \omega_{pl} < \omega$)

The plasma becomes a relatively low-loss dielectric. The attenuation is small for $\omega \gg \omega_{pl}$, but large for $\omega \le \omega_{pl}$. In this case

$$\mu_r \cong \left(1 - \frac{\omega_{pl}^2}{\omega^2}\right)^{1/2}$$

$$\mu_i \cong \frac{\nu\,\omega_{pl}^2}{2\,\omega^3} \left(1 - \frac{\omega_{pl}^2}{\omega^2}\right)^{-1/2}$$

$$\delta \cong \frac{c}{\omega_{pl}} \left(\frac{2\,\omega^2}{\nu\,\omega_{pl}} \right) \left(1 - \frac{\omega_{pl}^2}{\omega^2} \right)^{1/2}$$

$$v_\Phi = c \left(1 - \frac{\omega_{pl}^2}{\omega^2} \right)^{-1/2}$$

$$v_g = c \left(1 - \frac{\omega_{pl}^2}{\omega^2} \right)^{1/2}$$

Reference: [42]

8.2. PROPAGATION IN A LORENTZ PLASMA WITH AN EXTERNAL MAGNETIC FIELD

The electron equation of motion is

$$m_e \frac{\partial v}{\partial t} = -\,e \left(E(t) + v \times B_0 \right) - \nu\, m_e\, v$$

where B_0 is a static magnetic field. Waves propagating along B_0 are circularly polarized (cyclotron waves). The (complex) Lorentz conductivity is

$$\sigma_l = i \frac{n_e\, e^2}{m_e \left(i\,\nu - (\omega + \omega_c) \right)}$$

$$\sigma_r = i \frac{n_e\, e^2}{m_e \left(i\,\nu - (\omega - \omega_c) \right)}$$

where the subscripts l and r denote the left and the right circularly polarized wave, respectively; $\omega_c = |e|\, B_o/m_e$ is the → electron cyclotron frequency. In this case

* the (complex) Lorentz dielectric constant

$$\varepsilon_{l,r} = 1 - i \frac{\sigma_{l,r}}{\varepsilon_o\, \omega}$$

$$= 1 - \frac{\omega_{pl}^2}{\omega \left((\omega \pm \omega_c) - i\,\nu \right)}$$

$$\varepsilon_{l,r} = \left\{1 - \frac{\omega_{pl}^2 \ (\omega \pm \omega_c)}{\omega \left[(\omega \pm \omega_c)^2 + \nu^2\right]}\right\} - i \left\{\frac{\omega_{pl}^2 \ \nu}{\omega \left[(\omega \pm \omega_c)^2 + \nu^2\right]}\right\}$$

* the (real) <u>refractive index</u>

$$\mu_{l,r}^{(real)} = \mathcal{R}e \left[\left(\varepsilon_{l,r}\right)^{1/2}\right]$$

$$= \left[\frac{1}{2} \left\{1 - \frac{\omega_{pl}^2 \ (\omega \pm \omega_c)}{\omega \left[(\omega \pm \omega_c)^2 + \nu^2\right]}\right\} + \right.$$

$$\left. + \frac{1}{2} \left(\left\{ 1 - \frac{\omega_{pl}^2 \ (\omega \pm \omega_c)}{\omega \left[(\omega \pm \omega_c)^2 + \nu^2\right]}\right\}^2 + \left\{ \frac{\omega_{pl}^2 \ \nu}{\omega \left[(\omega \pm \omega_c)^2 + \nu^2\right]} \right\}^2 \right)^{1/2} \right]^{1/2}$$

* the <u>attenuation index</u>

$$\mu_{l,r}^{(imag)} = - \mathcal{I}m \left[\left(\varepsilon_{l,r}\right)^{1/2}\right]$$

$$= \left[- \frac{1}{2} \left\{1 - \frac{\omega_{pl}^2 \ (\omega \pm \omega_c)}{\omega \left[(\omega \pm \omega_c)^2 + \nu^2\right]}\right\} + \right.$$

$$\left. + \frac{1}{2} \left(\left\{ 1 - \frac{\omega_{pl}^2 \ (\omega \pm \omega_c)}{\omega \left[(\omega \pm \omega_c)^2 + \nu^2\right]}\right\}^2 + \left\{ \frac{\omega_{pl}^2 \ \nu}{\omega \left[(\omega \pm \omega_c)^2 + \nu^2\right]} \right\}^2 \right)^{1/2} \right]^{1/2}$$

If

$$\nu^2 \ll \begin{cases} (\omega \pm \omega_c)^2 \\ \\ |\omega \ (\omega \pm \omega_c) - \omega_{pl}^2| \end{cases}$$

then

$$\mu_{l,r}^{(real)} \cong \left\{ 1 - \frac{\omega_{pl}^2}{\omega \left[(\omega \pm \omega_c)^2\right]} \right\}^{1/2}$$

and

$$\mu_{l,r}^{(imag)} \cong \frac{\omega_{pl}^2 \ \nu}{2\omega\left(\omega \pm \omega_c\right)^{3/2}\left[\omega \pm \omega_c - \omega_{pl}^2/\omega\right]^{1/2}}$$

The refractive index, $\mu_{l,r}^{(real)}$, has a zero-point (cut-off condition) for

$$\frac{\omega_{pl}^2}{\omega^2} = 1 \pm \frac{\omega_c}{\omega}$$

and a pole (resonance condition) at the cyclotron resonance

$$\omega = \omega_c$$

for the right-hand circular polarized wave.

If a linearly polarized, plane wave is incident on a magnetized plasma along the field B_0, its direction of linear polarization will be rotated (Faraday rotation) through an angle

$$\varphi = \left(\mu_l^{(real)} - \mu_r^{(real)}\right) \pi \frac{d}{\lambda}$$

where d is the distance travelled in the plasma, and λ is the wavelength.

Reference: [42]

8.3. WEAKLY DAMPED WAVES IN A PLASMA IN THE ABSENCE OF AN EXTERNAL MAGNETIC FIELD

Nomenclature:

ω angular frequency of the wave

$\omega_{pl,e}$ → electron plasma frequency $\left[SI\text{: } \left(n_e e^2/\ \varepsilon_0\ m_e\right)^{1/2}\right]$

$\omega_{pl,i}$ → ion plasma frequency $\left[SI\text{: } \left(n_i e^2\ Z^2/\ \varepsilon_0\ m_i\right)^{1/2}\right]$

$\omega_{c,e}$ → electron cyclotron frequency $\left[SI\text{: } e\ B_0/\ m_e\right]$

$\omega_{c,i}$ → ion cyclotron frequency $\left[SI\text{: } Z\ e\ B_0/\ m_i\right]$

ω_{UH} → upper hybrid frequency (see 3.3.3.)

ω_{LH} → lower hybrid frequency (see 3.3.3.)

$\nu_e = \begin{cases} \nu_{eff} & \text{→ effective collision frequency for a completely ionized plasma} \\ \nu_{en} & \text{→ collision frequency of electrons with neutrals in a weakly ionized plasma} \end{cases}$

$\lambda_{D,e}$ → electron Debye length $\left[SI: \left(\varepsilon_0\, k_B T_e / n_e\, e^2\right)^{1/2}\right]$

k wave vector, $k = |\mathbf{k}| = 2\pi / \lambda$

k_z component of **k** in the z-direction (∥ **B**)

θ angle between **k** and B_0

k_B → Boltzmann's constant

$v_{T,e} = \left(2\, k_B T_e / m_e\right)^{1/2}$ electron thermal velocity

$v_{T,i} = \left(2\, k_B T_i / m_i\right)^{1/2}$ ion thermal velocity

v_A → Alfven velocity $\left[SI: B_0 / (\mu_0\, n_i\, m_i)^{1/2}\right]$

v_s → sound velocity $= \left\{\left(\gamma_e\, k_B T_e + \gamma_i\, k_B T_i\right) / m_i\right\}^{1/2}$, $\gamma_e \cong 1$, $\gamma_i \cong 3$

c → velocity of light in vacuum

type of waves	dispersion relation	conditions / remarks
electron Langmuir waves	$\omega^2 = \omega_{pl,e}^2 + \frac{3}{2} k^2\, v_{T,e}^2$ $= \omega_{pl,e}^2 \left(1 + 3\, k^2\, \lambda_{D,e}^2\right)$	$\omega \gg k\, v_{T,e}$, ν_e **k** ∥ **E**
ion Langmuir waves	$\omega^2 = \omega_{pl,i}^2 \dfrac{1 + 3\, k^2 \lambda_{D,i}^2 \;\; (1 + k^{-2} \lambda_{D,e}^{-2})}{1 + k^{-2}\, \lambda_{D,e}^{-2}}$ $\cong \omega_{pl,i}^2 \dfrac{1}{1 + k^{-2}\, \lambda_{D,e}^{-2}}$	$k\, v_{T,i} \ll \omega \ll k\, v_{T,e}$ **k** ∥ **E**
ion sound waves	$\omega^2 = k^2\, v_s^2$ $= k^2 \left(\gamma_e\, k_B T_e + \gamma_i\, k_B T_i\right) / m_i$	$k\, v_{T,i} \ll \omega \ll k\, v_{T,e}$ $k\, \lambda_{D,e} \ll 1$ **k** ∥ **E**
transverse electromagnetic waves	$\omega^2 = \omega_{pl,e}^2 + k^2\, c^2$	$\omega \gg k\, v_{T,e}$, ν_e **k** ⊥ **E**

8.4. WEAKLY DAMPED WAVES IN A COLD PLASMA WITH AN EXTERNAL MAGNETIC FIELD

Notation as in section 8.3.

In a "cold" plasma the relation

$$\left.\begin{array}{ll} SI: & n_e\, k_B T_e / 2\,\mu_0\, B_0^2 \\ \\ CGS: & 8\pi\, n_e\, k_B T_e / B_0^2 \end{array}\right\} \ll 1$$

applies. The phase velocity is faster than the electron and ion thermal velocities. For $\omega \gg \omega_{c,e}$, the magnetic field is not important and the spectra will be identical to those in its absence.

type of waves	dispersion relation	conditions / remarks
longitudinal waves	$\omega_1^2 = \frac{1}{2}\left(\omega_{pl,e}^2 + \omega_{c,e}^2\right) + \frac{1}{2}\left(\left[\omega_{pl,e}^2 + \omega_{c,e}^2\right]^2 - 4\,\omega_{pl,e}^2\,\omega_{c,e}^2 \cos^2\theta\right)^{1/2}$	
	for $\cos\theta \geq m_e/m_i$ and $\omega \gg k_z\, v_{T,\alpha}$ $(\alpha = e,\ i)$:	
	$\omega_2^2 = \frac{1}{2}\left(\omega_{pl,e}^2 + \omega_{c,e}^2\right) - \frac{1}{2}\left(\left[\omega_{pl,e}^2 + \omega_{c,e}^2\right]^2 - 4\,\omega_{pl,e}^2\,\omega_{c,e}^2 \cos^2\theta\right)^{1/2}$	
	$\omega_3 = \omega_{c,i}\left(1 - \frac{m_e}{m_i}\tan^2\theta\right)$	
	for $\cos^2\theta \ll m_e/m_i$	
	$\omega_2^2 = \omega_{c,i}^2 + \frac{\omega_{c,e}^2\ \omega_{pl,i}^2}{\omega_{c,e}^2 + \omega_{pl,e}^2} \equiv \omega_{LH}^2$	
	$\omega_3^2 = \frac{\omega_{c,i}^2\ \omega_{pl,e}^2}{\omega_{c,e}^2 + \omega_{pl,e}^2} \cos^2\theta$	
	$\omega_1 \longrightarrow max\left(\omega_{c,e},\ \omega_{pl,e}\right)$	for $\theta \longrightarrow 0$,
	$\omega_2 \longrightarrow min\left(\omega_{c,e},\ \omega_{pl,e}\right)$	i.e. $\mathbf{k} \parallel \mathbf{B}$
(continued)	$\omega_3 \longrightarrow \omega_{c,i}$	

type of waves	dispersion relation	conditions / remarks
longitudinal waves	$\omega_1 \longrightarrow \left(\omega^2_{pl,e} + \omega^2_{c,e}\right)^{1/2} \equiv \omega_{UH}$ $\omega_2 \longrightarrow \omega_{LH}$ $\omega_3 \longrightarrow 0$	for $\theta \longrightarrow \pi/2$, i.e. $k \perp B$
Alfven wave	$\omega^2 = \dfrac{k^2\, v_A^2 \cos\theta}{1 + v_A^2/c^2}$ $\omega^2 = k^2\, v_A^2$	$k_z\, v_{T,i} \ll \omega \ll \omega_{c,i}$ $k \parallel B_0$ and $v_A^2 \ll c^2$
compression Alfvén waves	$\omega^2 = \dfrac{k\, v_A}{1 + v_A^2/c^2}$ electromagnetically transverse, but mechanically longitudinal waves	$k_z\, v_{T,i} \ll \omega \ll \omega_{c,i}$
spiral waves (whistler; in solids helicon waves)	$\omega = \lvert\omega_{c,e} \cos\theta\rvert \dfrac{k^2\, c^2}{\omega^2_{pl,e}}$	$k_z v_{T,\alpha}, \omega_{c,i} \ll \omega \ll \omega_{c,e}$ $\omega\, \omega_{c,e} \ll \omega^2_{pl,e}$ $\lvert 1 - \cos^2\theta\rvert \ll 1$

special cases of electromagnetic waves: $k \perp B_0$

light waves ordinary wave	$\omega^2 = \omega^2_{pl,e} + k^2\, c^2$ or alternatively $\dfrac{c^2\, k^2}{\omega^2} = 1 - \dfrac{\omega^2_{pl,e}}{\omega^2}$	$E(t) \parallel B_0$
extraordinary waves	$\dfrac{c^2\, k^2}{\omega^2} = 1 - \dfrac{\omega^2_{pl,e}}{\omega^2}\left(\dfrac{\omega^2 - \omega^2_{pl,e}}{\omega^2 - \omega^2_{UH}}\right)$ $\omega_{UH} \rightarrow$ upper hybrid frequency	$E(t) \perp B_0$

type of waves	dispersion relation	conditions / remarks
		$k \perp B_0$
left-hand circularly polarized wave	$\frac{c^2 k^2}{\omega^2} = 1 - \frac{\omega_{pl,e}^2}{\omega^2} \left(\frac{1}{1 + \omega_{c,e}/\omega} \right)$	
right-hand circularly polarized wave	$\frac{c^2 k^2}{\omega^2} = 1 - \frac{\omega_{pl,e}^2}{\omega^2} \left(\frac{1}{1 - \omega_{c,e}/\omega} \right)$	

8.5. WEAKLY DAMPED WAVES IN A WARM PLASMA WITH AN EXTERNAL MAGNETIC FIELD

Notation as in section 8.3.

In a "warm" plasma, the electron thermal motion cannot be neglected. Collisions (or a non-zero pressure) must be considered. Collisional damping and new classes of waves appear.

Longitudinal waves ($k \parallel B_0$) are not influenced by the magnetic field, the respective waves may be found in section 8.3.

type of waves	dispersion relation	conditions / remarks
ion sound waves	$\omega^2 = \frac{k^2 v_s^2 \cos^2\theta}{1 + k^2 \lambda_{D,e}^2}$ $\omega^2 = \frac{k^2 v_s^2}{1 + k^2 \lambda_{D,e}^2}$	$k_z v_{T,i} < \omega < k_z v_{T,e}$ $\omega \ll \omega_{c,e}$; $\omega_{c,i} \ll \omega \ll \omega_{c,e}$
Alfvén wave	$\omega^2 = k^2 v_A^2 \cos\theta$	$\nu_i \ll \omega \ll \omega_{c,i}$ $\nu_e \ll \omega_{c,e}$ $v_{T,i} \ll v_A \ll v_{T,e}$

type of waves	dispersion relation	conditions / remarks
		$k_z v_{T,i} \ll \omega \ll k_z v_{T,e}$ $\omega \ll \omega_{c,i}$ $k_\perp^2 \ll v_{T,i}^2 / \omega_{c,i}^2$
fast magneto-sound waves	$\omega_+^2 = \frac{1}{2} k^2 \left[v_A^2 + v_s^2 + \left(\left(v_A^2 + v_s^2 \right)^2 - 4 v_A^2 v_s^2 \cos^2\theta \right)^{1/2} \right]$	
slow magneto-sound waves	$\omega_-^2 = \frac{1}{2} k^2 \left[v_A^2 + v_s^2 - \left(\left(v_A^2 + v_s^2 \right)^2 - 4 v_A^2 v_s^2 \cos^2\theta \right)^{1/2} \right]$ In the case of a low-pressure plasma, when $v_A^2 \gg v_s^2$, $\omega_+^2 \cong k^2 v_A^2$ is the continuation of the compression Alfvén wave into the region of lower frequencies $\omega_-^2 \cong k^2 v_s^2$ is the continuation of the ion-sound wave for the case of a magnetized plasma in addition to a sound wave in a non-isothermal plasma	
spiral waves	$\omega = \lvert \omega_{c,e} \cos\theta \rvert \dfrac{k^2 c^2}{\omega_{pl,e}^2}$	$k_z v_{T,i} \ll \omega \ll k_z v_{T,e}$ $\omega_{c,i} \ll \omega \ll \omega_{c,e}$
cyclotron waves	$\omega \cong N \omega_{c,e \text{ or } i}$, $N = 1,2,3,\ldots$ $\omega \cong \omega_{c,e/i} \left(1 - \dfrac{\omega_{pl,e/i}^2}{k^2 c^2} \right)$	weakly damped only outside the resonace line, i.e. $\lvert \omega - \omega_{c,e/i} \rvert \gg k_z v_{T,e/i}$

Resonance: phase velocity $v_\Phi \longrightarrow 0$, refractive index $\mu = \dfrac{c\,k}{\omega} = \dfrac{c}{v_\Phi} \longrightarrow \infty$,
$k \longrightarrow \infty$, $\lambda \longrightarrow 0$, pole of the dispersion relation.

Cut-off: phase velocity $v_\Phi \longrightarrow \infty$, refractive index $\mu \longrightarrow 0$,
$k \longrightarrow 0$, $\lambda \longrightarrow \infty$, zero of the disperion relation.

In both cases, the group velocity will be zero, $v_g \longrightarrow 0$. Propagating waves will be absorbed or reflected at points of resonance or cut-off respectively. This is shown in the figure below (Clemmow-Mullaly-Allis diagram, [37] vol.II p.131, [42]):

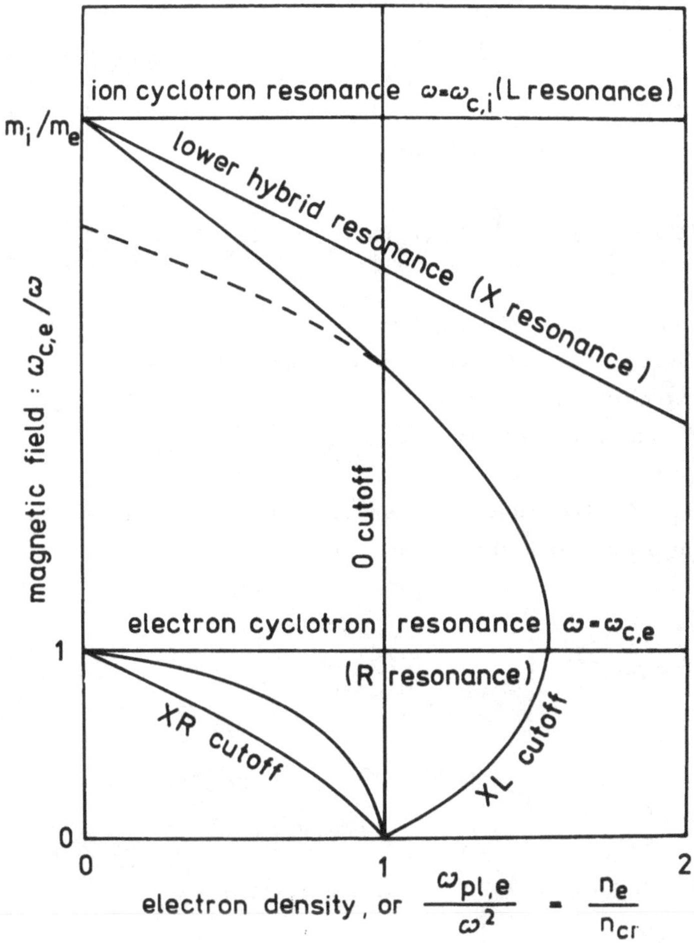

References: [37], [42], [60], [75], [103]

9. ELEMENTARY PROCESSES

9.1. ELECTRON IMPACT EXCITATION

The transition rate due to electron impact may be calculated by two widely divergent methods:

(i) the extreme quantum limit (i.e. Born approximation)

(ii) classical binary-collision method

The rate coefficient of electron impact excitation for a transition from a lower state l to an upper state u is defined by

$$C_{lu}^{(e)} = \langle \sigma_{lu}^{(e)} v_e \rangle = \int \sigma_{lu}^{(e)}(v_e)\, f(v_e)\, v_e\, dv_e$$

where

$\sigma_{lu}^{(e)}$: cross-section for the excitation from l to u

v_e : relative velocity of the colliding particles (can be assumed equal to the electron velocity)

$f(v_e)$: electron distribution function

Using → Drawin's formula for the excitation cross section and a → Maxwellian distribution function, the rate coefficient becomes [9]

$$C_{lu}^{(e)} = C\, f_{lu} \frac{u_H^{3/2}}{u_{lu}} \begin{cases} \Psi_1(u_{lu}, \beta) & \text{for neutrals} \\ \Psi_2(u_{lu}, \beta) & \text{for ions} \end{cases}$$

where

$$C = K_\varepsilon^{-1/2}\, e\, a_0^{3/2}\, m_e^{-1/2} = \begin{cases} 8.686 \times 10^{-14}\ \mathrm{m^3/s} & SI \\ 8.868 \times 10^{-8}\ \mathrm{cm^3/s} & CGS \end{cases}$$

f_{lu} : → absorption oscillator strength

$u_H = E_H/kT$ ($E_H = 13.6$ eV)

$u_{lu} = |E_u - E_l|/kT$ (normalized energy difference between l and u)

Ψ_1, Ψ_2 : quantum mechanical correction factors to the classical result ([9], [89])

$$\beta = \begin{cases} 1 & \text{for ions} \\ 2 & \text{for neutrals} \end{cases}$$

Drawin's result cooresponds to Born approximation calculations at high energies (Bethe form) [196]:

$$C_{lu}^{(e)} = 16\,\pi \left(\frac{2}{3}\,\pi\right)^{1/2} \frac{a_0^2\; E_H^2\; f_{lu}}{(m_e\; kT)^{1/2}\; |E_u - E_l|}\; exp\left(-\,u_{ul}\right)\, G_{ff}$$

where

$$G_{ff} \cong \frac{3^{1/2}}{\pi}\; \ell n\left(\frac{4\; E_e}{|E_u - E_l|}\right)$$

is the free-free electron Gaunt factor.

From classical considerations, when the kinetic energy of the free electron is large compared with $|E_u - E_l|$,

$$C_{lu}^{(e)} = K_\varepsilon^2\; 2\pi\; e^4 \left(\frac{2}{\pi\; m_e\; kT}\right)^{1/2} \frac{\Delta E_u}{|E_u - E_l|}\; exp\left(-\,u_{ul}\right) \left\{\left(1 + \frac{4}{3}\,\frac{E_l}{|E_u - E_l|}\right)\right.$$

$$\left. + \left(\frac{5}{3} - \frac{2}{3}\,\frac{E_l}{kT}\right) \frac{|E_u - E_l|^2}{E_l\; kT} - \left(u_{ul}^2 + u_{ul} - \frac{2}{3}\,\frac{|E_u - E_l|^2}{(kT)^3}\,E_l\right) exp\left(\frac{E_l}{kT}\right)\, \mathcal{E}i\left(\frac{E_l}{kT}\right)\right\}$$

where

$$\Delta E_u \cong \frac{2\; Z_{eff}^2\; E_H}{u^3}$$

is the energy width of the upper state u, $\mathcal{E}i$ is the exponential integral, and E_l denotes the ionization energy of the lower state.

References: [9]-[13], [24], [32], [33], [178], [179], [196]

9.2. ELECTRON IMPACT DE-EXCITATION

Using the → rate coefficient $C_{lu}^{(e)}$ of excitation and the method of detailed balance, one obtains the de-excitation rate coefficient due to electron impact

$$F_{ul}^{(e)} = \langle\sigma_{ul}^{(e)}\; v_e\rangle = \frac{g_l}{g_u}\; exp\left(u_{lu}\right)\; C_{lu}^{(e)}$$

where g_l, g_u are the → statistical weigths of the lower and the upper level respectively.

Reference: [9], [196]

9.3. ELECTRON IMPACT IONIZATION

The rate coefficient is

$$S_n^{(e)} = \langle\sigma_{in}^{(e)} v_e\rangle = \int \sigma_{in}^{(e)}(v_e)\, f(v_e)\, v_e\, dv_e$$

where $\sigma_{in}^{(e)}$ is the → cross-section for ionization from the *n*-th level by electron impact, the other symbols were introduced in section 9.1.
Using Drawin's formula for $\sigma_{in}^{(e)}$, one obtains

$$S_n^{(e)} = 8\,\pi\, a_0^2 \left(\frac{E_H}{E_n}\right)^2 \xi_n\, f_n \left(\frac{2\,k\,T_e}{\pi\, m_e}\right)^{1/2} u_n\, \Psi_1(u_n, \beta)$$

with

$u_n = E_n/kT_e$ normalized binding energy of electrons in level *n* and

$$\Psi_1(u_n, \beta) = \int_{u_n}^{\infty} \left\{\left(1 - u_n/U\right) \exp(-U)\, \ln\left(1.25\,\beta\, U/u_n\right)\right\} dU$$

Meanings of all other symbols are explained in section 3.2.7. covering ionization cross-sections.
References: [9]-[26], [32], [33], [179]

9.4. ELECTRONIC RECOMBINATION

9.4.1. THREE-PARTICLE RECOMBINATION IN A HIGHLY IONIZED PLASMA

In a dense, low-temperature plasma, recombination is mainly governed by three-particle recombination $X^+ + e^- + e^- \longrightarrow X + e^-$ ($dn_e/dt \sim \alpha_{coll}\, n_i\, n_e^2$). Using the → rate coefficient, $S_n^{(e)}$, for electron impact ionization from the *n*-th level and the method of detailed balance, the coefficient for three-particle recombination into the level *n* is given by [9]

$$\alpha_n^{(e)} = \frac{g_n}{2\, g_z} \frac{h^3}{(2\pi\, m_e\, k\, T_e)^{3/2}} \exp(u_n)\, S_n^{(e)}$$

where

g_n : statistical weight of the level *n*

g_z : statistical weight of the ground state of the ion

$u_n = E_n/kT_e$ normalized binding energy of electrons in the level *n*.

The total rate coefficient can be explicitly expressed in an approximation which is valid for three-particle recombination in a dense, low-temperature plasma [105]:

$$\alpha_{coll} = K_\varepsilon^5 \frac{4\ 2^{1/2}\ \pi^{3/2}\ e^{10}\ Z^3\ \ell n\,[(Z^2+1)^{1/2}]}{9\ \ m_e^{1/2}\ \ (k\ T_e)^{9/2}}$$

$$= \begin{cases} 1.78 \times 10^{-20}\ Z^3\ \ell n\,[(Z^2+1)^{1/2}]\ T_e^{-9/2} & (SI) \\ 1.78 \times 10^{-8}\ \ Z^3\ \ell n\,[(Z^2+1)^{1/2}]\ T^{-9/2} & (CGS) \end{cases}$$

$$\cong \begin{cases} 6.2 \times 10^{-21}\ T_e^{-9/2} & (SI) \\ 6.2 \times 10^{-9}\ \ T^{-9/2} & (CGS) \end{cases} \qquad \text{for } Z = 1;$$

	α_{coll}	T_e
SI	m^6/s	K
CGS	cm^6/s	K

where Z is the mean ion charge. For $Z \gg 1$, $\ell n\left((Z^2+1)^{1/2}\right) \longrightarrow \ell n\ \Lambda$ (⟶ Coulomb logarithm).

An approximate formula is [33]

$$\alpha_{coll} \cong \begin{cases} 5.4 \times 10^{-27}\ \ T_e^{-9/2} & (\alpha \text{ in } cm^6/s,\ T_e \text{ in eV}) \\ 1.1 \times 10^{-39}\ \ T_e^{-9/2} & (\alpha \text{ in } m^6/s,\ T_e \text{ in eV}) \end{cases}$$

<u>References</u>: [9]-[13], [33], [105]

9.4.2. THREE-PARTICLE RECOMBINATION IN A WEAKLY IONIZED PLASMA

A (neutral) gas particle may take part in the recombination process $X^+ + e^- + X \longrightarrow X + X$ $(dn_e/dt \sim \alpha_{coll}\ n_i\ n_e\ n_a)$. The corresponding recombination coefficient can be written as [106]

$$\alpha_{coll} = K_\varepsilon^3 \frac{32\ (2\pi)^{1/2}\ m_e^{1/2}\ e^6\ \sigma_{ea}}{3\ \ m_a\ \ (k\ T)^{5/2}}$$

$$= \begin{cases} 4.42 \times 10^{-40}\ \sigma_{ea}\ /\ (m_a\ T^{5/2}) = 2.66 \times 10^{-13}\ \sigma_{ea}\ /\ (M_a\ T^{5/2}) & (SI) \\ 4.42 \times 10^{-29}\ \sigma_{ea}\ /\ (m_a\ T^{5/2}) = 2.66 \times 10^{-5}\ \ \sigma_{ea}\ /\ (M_a\ T^{5/2}) & (CGS) \end{cases}$$

	α_{coll}	σ_{ea}	n_a	m_a	M_a	T
SI	m^6/s	m^2	m^{-3}	kg	-	K
CGS	cm^6/s	cm^2	cm^{-3}	g	-	K

where

σ_{ea} : effective cross-section for electron-atom interaction

n_a : atom number density

m_a : atomic mass (M_a is the atomic mass number in → units m_u)

T : temperature

The formula is valid if

$$kT \ll \left(m_e / m_a\right)^{1/2} K_\varepsilon \frac{e^2}{a}$$

"a" is the characteristic radius of the atom, that is for helium: T < 1500 K.

Reference: [106]

9.4.3. DISSOCIATIVE RECOMBINATION

$$dn_e / dt \sim n_e\, n_2^+\, \alpha_{diss}$$

where n_2^+ denotes the molecular ion density. The coefficient of dissociative recombination can be expressed as

$$\alpha'_{diss} \cong \frac{g_{auto}}{2\,\sigma_i} \frac{exp\left(-E_{auto} / kT_e\right) h^3}{(2\pi\, m_e\, k\, T_e)^{3/2}\, a_0} \left(\frac{k\, T_i}{m_i}\right)^{1/2}$$

where

g_{auto} : statistical weight of the autoionization state

E_{auto} : excitation energy of the autoionization state

σ_i : → partition function of the ion

a_0 : characteristic atomic length (→ Bohr radius)

Typically, $\alpha'_{diss} \cong 10^{-14} - 10^{-12}\ m^3/s$.

Reference: [33]

9.4.4. RADIATIVE RECOMBINATION

$$dn_e/dt \sim n_i\, n_e\, \alpha'_{rad}$$

The coefficient of radiative recombination is given by

$$\alpha'_{rad} = \sum_k A_{ek} \sim 10^{-19} - 10^{-18}\ m^3/s$$

The transition rate (electron continuum $\longrightarrow$ level *k*) can be estimated with

$$A_{ek} \cong K_\varepsilon \, \alpha_F^3 \, a_o^2 \, e^2 / \hbar$$

where

α_F : fine structure constant (1/137)

a_o : Bohr radius

A more detailed consideration gives

$$A_{ek} = \int f_e(\mathcal{E}) \left(\frac{2\,\mathcal{E}}{m_e}\right)^{1/2} \mathcal{E}^{1/2} \, \sigma^{ph.rec.}_{e\rightarrow k}(\mathcal{E}) \, d\mathcal{E}$$

$$\cong \underbrace{\frac{64\,\pi^{1/2} a_0 \,\hbar\, \alpha^3}{3\;3^{1/2}\; m_e}}_{\text{const}} Z \left(\frac{Z^2\,Ry}{k^2\,kT_e}\right)^{3/2} \mathit{exp}\left(\frac{Z^2\,Ry}{k^2\,kT_e}\right) \mathcal{E}i\left(\frac{Z^2\,Ry}{k^2\,kT_e}\right)$$

$$\text{const} = \begin{cases} 5.20 \times 10^{-20}\ \mathrm{m^3/s} & (SI) \\ 5.20 \times 10^{-14}\ \mathrm{cm^3/s} & (CGS) \end{cases}$$

where

$\sigma^{ph.rec.}_{e\rightarrow k}(\mathcal{E})$: $\rightarrow$ cross-section for photo-recombination

$f_e(\mathcal{E})$: electron distribution function

$\mathcal{E}i(\;)$: integral exponent

Integrating over the *velocity* distribution function, an equivalent expression is obtained

$$\alpha'_{rad} = \sum_k \langle \sigma^{ph.rec.}_{e\rightarrow k}(v_e)\, v_e \rangle = \sum_k \int_0^\infty \sigma^{ph.rec.}_{e\rightarrow k}(v_e)\, f_e(v_e)\, v_e\, dv_e$$

Using the $\rightarrow$ cross-section for photo-recombination and assuming a $\rightarrow$ Maxwellian velocity distribution, the coefficient α'_{rad} can be written

$$\alpha'_{rad}(T_e) = A_{ph.rec.} \left(\frac{8\,kT_e}{\pi\, m_e}\right)^{1/2} C\,\Phi(C) = \begin{cases} 2.065 \times 10^{-17}\, Z^2\, T_e^{-1/2}\, \Phi(C) & (SI) \\ 2.065 \times 10^{-11}\, Z^2\, T_e^{-1/2}\, \Phi(C) & (CGS) \end{cases}$$

where

$A_{ph.rec.}$ is the constant of electron capture ($\rightarrow$ cross-section for photo-recombination)

$$C = \frac{h\,\nu_0}{kT_e} = \frac{Z^2\,Ry}{kT_e} = 157880\text{ K}\,\frac{Z^2}{T_e} \quad \text{(Z is the mean ion charge, } n_e = Z\,n_i\text{)}$$

$$\phi(C) = \sum_{k=1}^{\infty} \frac{C}{k^3}\, exp\left(\frac{C}{k^2}\right)\left\{-\,\mathcal{E}i\left(-\frac{C}{k^2}\right)\right\}$$

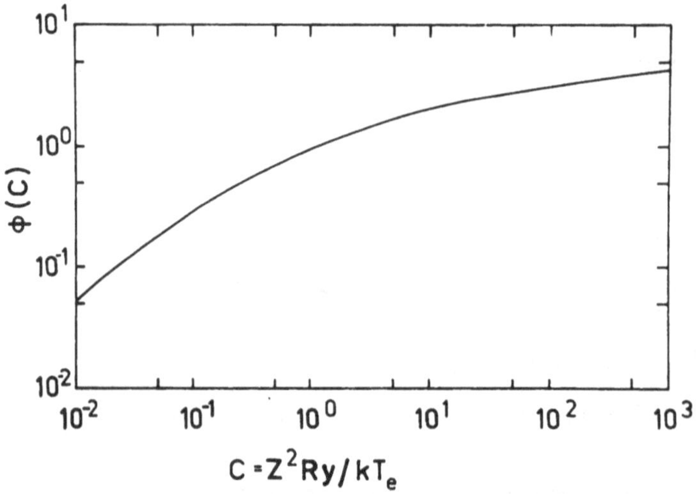

	α'_{rad}	$A_{ph.rec.}$	C	$\phi(C)$	T_e	Z
SI	m^3/s	m^2	-	-	K	-
CGS	cm^3/s	cm^2	-	-	K	-

References: [9], [31] p.196, [33]

9.5. RADIATIVE DECAY

The radiation decay probability from an upper state *u* to a lower state *l* is given by

$$A_{ul} = K_\varepsilon \frac{8\,\pi^2\,e^2\,|E_u - E_l|^2}{m_e\,h^2\,c^3}\,\frac{g_l}{g_u}\,f_{lu} = 4.34 \times 10^7\,|E_u - E_l|^2\,\frac{g_l}{g_u}\,f_{lu}$$

	A_{ul}	$g_{u/l}$	f_{lu}	$\lvert E_u - E_l\rvert$	$\lvert E_u - E_l\rvert$ in the last (numerical) term
SI	s^{-1}	-	-	J	eV
CGS	s^{-1}	-	-	erg	eV

See also → Ladenburg formula in section 10.4.2.

10. RADIATION

10.1. GENERAL

Each photon is characterized by

(1) an energy E (or frequency ν, where $E = h\nu$, or wavelength λ, where $E = hc/\lambda$)

(2) three components of the radius vector r, which specifies the location of the photon in space

(3) two angular variables giving the direction of motion (direction cosines)

(This is valid in the absence of polarization. The case of polarization is discussed in [47].)

An ensemble of photons (a radiation field) is described by a *distribution function* $f(\nu,r,\Omega,t)$, such that $f(\nu,r,\Omega,t)\ d\nu\ dV\ d\Omega$ is the number of photons with frequencies between ν and $\nu+d\nu$ at time t in a volume dV near a point r and moving within a solid angle $d\Omega$ around the direction Ω. If f does not depend on time, the radiation field is said to be steady-state.

The monochromatic intensity of radiation $L_\nu(r,\Omega)$ is defined through the relation

$$L_\nu(r,\Omega) = c\ h\nu\ f(\nu,r,\Omega)$$

	L_ν	$f(\nu,r,\Omega)$
SI	J/s m^2 sr Hz	1/m^3 Hz sr
CGS	erg/s cm^2 sr Hz	1/cm^3 Hz sr

It is also referred to as radiance; To avoid confusion it is advisable to check the units.

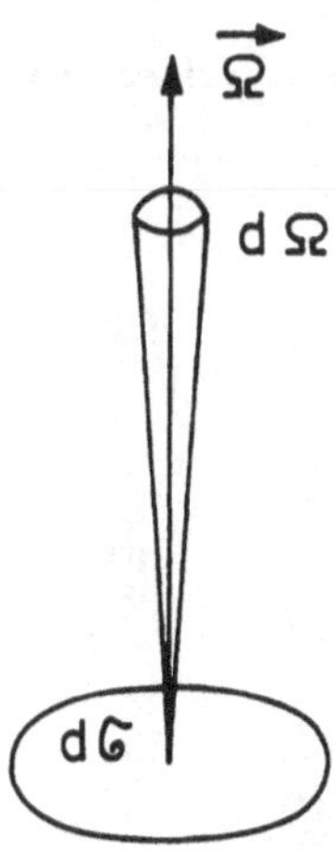

$L_\nu(r,\Omega)\ d\nu\ d\sigma\ d\Omega$ represents the energy transferd by photons with frequencies between ν and $\nu+d\nu$ per unit time through the area $d\sigma$ in directions within the solid angle $d\Omega$ located at the point r.

Using

$$c = \nu\ \lambda$$

and

$$|d\nu| = \frac{c}{\lambda^2}\ |d\lambda|$$

all monochromatic quantities related to the frequency ν can be transformed into quantities related to the wavelength λ,

$$L_\lambda = \frac{c}{\lambda^2} L_\nu$$

or

$$\nu L_\nu = \lambda L_\lambda$$

Integration over frequency (or wavelength) gives the total intensity

$$L = \int_0^\infty L_\nu \, d\nu = \int_0^\infty L_\lambda \, d\lambda$$

	L	L_ν	L_λ
SI	J/s m^2 sr	J/s m^2 sr Hz	J/s m^2 sr m
CGS	erg/s cm^2 sr	erg/s cm^2 sr Hz	erg/s cm^2 sr cm

The mean monochromatic intensity is defined as

$$\overline{L}_\nu(r) = \frac{1}{4\pi} \int L_\nu(r,\Omega') \, d\Omega'$$

Another integral over the distribution function f characterizing the radiation field is the monochromatic radiation density

$$\rho_\nu(r) = h\nu \int f(\nu,r,\Omega') \, d\Omega'$$

$$= \frac{1}{c} \int L_\nu(r,\Omega') \, d\Omega' = \frac{4\pi}{c} \overline{L}_\nu(r)$$

Moreover, the radiation field at a given point, r, is characterized by the monochromatic radiation flux vector $\pi F_\nu(r)$, defined by

$$\pi F_\nu(r) = \int \Omega' L_\nu(r,\Omega') \, d\Omega'$$

	$\overline{L}_\nu$	ρ_ν	πF_ν
SI	J/s m^2 Hz	J/m^3 Hz	J/s m^2 Hz
CGS	erg/s cm^2 Hz	erg/cm^3 Hz	erg/s cm^2 Hz

The projection of the flux vector on a given direction Ω is called the flux in the direction Ω and is denoted by

$$\pi F_{\nu,\Omega}(r) = \int L_\nu(r,\Omega') \cos\vartheta \, d\Omega'$$

With $\cos\vartheta = \Omega' \circ \Omega$

$$\pi F_{\nu,\Omega}(r) = \int_{\varphi=0}^{\pi} \int_{\vartheta=0}^{2\pi} L_\nu(r,\vartheta,\varphi) \cos\vartheta \sin\vartheta \, d\vartheta \, d\varphi$$

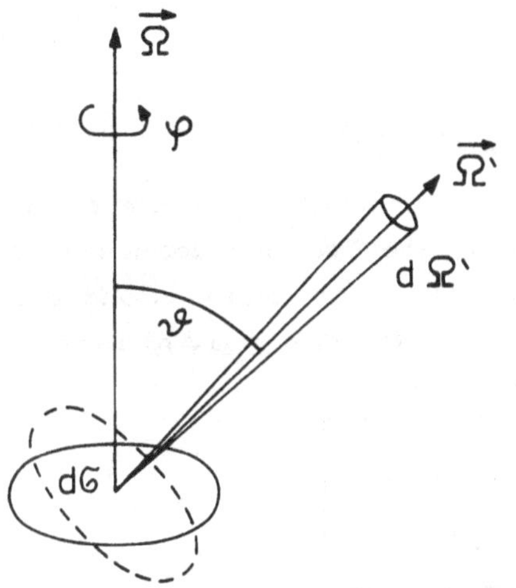

$\pi F_{\nu,\Omega}(r) \; d\sigma$ is the energy flowing across the area $d\sigma$ perpendicular to Ω, per unit time in the frequency interval $(\nu,\nu+d\nu)$. Often, it is expedient to divide $\pi F_{\nu,\Omega}$ into two parts:

(i) emittance

$(0 \le \vartheta \le \pi/2)$

$$\pi F^{+}_{\nu,\Omega} = \int_{0}^{2\pi} \int_{0}^{\pi/2} L_\nu \cos\vartheta \sin\vartheta \, d\vartheta \, d\varphi$$

(ii) irradiance

$(\pi/2 \le \vartheta \le \pi)$

$$\pi F^{-}_{\nu,\Omega} = \int_{0}^{2\pi} \int_{\pi}^{\pi/2} L_\nu \cos\vartheta \sin\vartheta \, d\vartheta \, d\varphi$$

with

$$F_{\nu,\Omega} = F^{+}_{\nu,\Omega} - F^{-}_{\nu,\Omega}$$

If the radiation field is isotropic, then $\pi F_{\nu,\Omega} = 0$. By analogy with the definition of L, the <u>total</u> <u>radiation</u> <u>flux</u> <u>vector</u> can be defined as

$$\pi F = \int_0^\infty \pi F_\nu \, d\nu = \int_0^\infty \pi F_\lambda \, d\lambda$$

	πF
SI	J/s m^2
CGS	erg/s cm^2

Reference: [48], [54]

10.2. RADIATION IN THERMODYNAMIC EQUILIBRIUM

In TE the radiation intensity has no spatial or directional dependence so that the radiation field is homogeneous and isotropic. The frequency dependence of the intensity is a universal function of the system temperature, T, and does not depend upon the properties of matter: this is the Planck law of radiation (also: Planck function or Kirchhoff-Planck function)

$$B_\nu(T) = \frac{2 h \nu^3}{c^2} \frac{1}{\exp\left(\frac{h\nu}{kT}\right) - 1}$$

$B_\nu d\nu$ is the energy emitted per unit time, surface area and solid angle in the frequency interval $(\nu, \nu+d\nu)$. With the relations $B_\nu d\nu = B_\lambda d\lambda$ and $|d\nu| = (c/\lambda^2)\ |d\lambda|$, the Planck law can be rewritten as

$$B_\lambda(T) = \frac{2 h c^2}{\lambda^5} \frac{1}{\exp\left(\frac{h c}{\lambda kT}\right) - 1}$$

	B_ν	B_λ	ν	λ	T
SI	J/s m^2 sr Hz	J/s m^2 sr m	s^{-1}	m	K
CGS	erg/s cm^2 sr Hz	erg/s cm^2 sr Hz	s^{-1}	cm	K

In the short wavelength region, $h\nu \gg kT$, the Planck law reduces to the Wien law of radiation

$$B_\nu(T) = \frac{2 h \nu^3}{c^2} \exp\left(- \frac{h\nu}{kT}\right)$$

Note, the substitution of Wien's law for Planck's law is equivalent to the assumption that stimulated emission is negligible in comparison with spontaneous emission ([48] p.4).

In the long wavelength region, $h\nu \ll kT$, the Planck law reduces to the Raleigh-Jeans law of radiation

$$B_\nu(T) = \frac{2\,\nu^2\,kT}{c^2}$$

Note, $h\nu \ll kT$ is the classical case in quantum mechanics (Bohr's correspondence principle) and therefore the Planck constant h does not occur here.

Wien displacement law

$$\lambda_{max}\,kT = hc/4.96511423$$

or

$$\lambda_{max}\,T = c_2/4.96511423$$

(numerical value from [1])

or

$$\lambda_{max} = 2.897756 \times 10^6\ T^{-1}$$

(T in K, λ_{max} in nm), where

λ_{max} : wavelength of maximum black body emission
T : temperature of the black body
c_2 : → second radiation constant

Stefan-Boltzmann law of radiation

$$\pi\,B(T) = \sigma_{St/B}\,T^4$$

where $B(T) = \int_0^\infty B_\nu(T)\,d\nu$, B_ν ⟶ Planck function

	B
SI	W/m^2 sr
CGS	$erg/s\ cm^2$ sr

π B(T) is the energy emitted per unit time, per unit surface area of a black body into a hemisphere surrounding the black body. The Stefan-Boltzmann constant is given

$$\sigma_{St/B} = \frac{2\,\pi^5\,k^4}{15\,c^2\,h^3} = \begin{cases} 5.6703 \times 10^{-8}\ W/m^2\ K^4 & (SI) \\ 5.6703 \times 10^{-5}\ erg/s\ cm^2\ K^4 & (CGS) \end{cases}$$

It is convenient to introduce the special notation, $u_\nu(T)$, for the monochromatic radiation density in thermodynamic equilibrium (TE). We have, from the definitions of ρ_ν and B_ν,

$$u_\nu(T) = \frac{8 \pi h \nu^3}{c^3} \frac{1}{\exp\left(\frac{h\nu}{kT}\right) - 1}$$

The integral radiation density

$$u(T) = \int_0^\infty u_\nu(T)\, d\nu$$

is proportional to the fourth power of the temperature

$$u(T) = a\, T^4$$

	u_ν	u
SI	J/m^3 Hz	J/m^3
CGS	erg/cm^3 Hz	erg/cm^3

where

$$a = \frac{8 \pi^5 k^4}{15 h^3 c^3} = \frac{4}{c} \sigma_{St/B} = \begin{cases} 7.566 \times 10^{-16}\ J/m^3\ K^4 & (SI) \\ 7.566 \times 10^{-15}\ erg/cm^3\ K^4 & (CGS) \end{cases}$$

Kirchhoff law of radiation

In TE, the ratio of the emission coefficient, ε_ν, to the absorption coefficient, $\varkappa_\nu$, is a universal function depending only on ν and T. This function gives the intensity of an ideal black body radiator

$$\frac{\varepsilon_\nu}{\varkappa_\nu} = B_\nu(T)$$

The law may also expressed as follows:
In TE, the ratio of the emissivity E_ν to the absorptance A_ν equals the intensity of the cavity radiation B_ν

$$\frac{E_\nu}{A_\nu} = B_\nu(T)$$

	B_ν	ε_ν	$\varkappa_\nu$	E_ν	A_ν
SI	W/m^2 sr Hz	W/m^3 sr Hz	m^{-1}	W/m^2 sr Hz	-
CGS	erg/s cm^2 sr Hz	erg/s cm^3 sr Hz	cm^{-1}	erg/s cm^2 sr Hz	-

The thermal emission of a plasma sheath is

$$L_\nu(T,z) = B_\nu(T)\left[1 - \exp(-\tau_\nu)\right] \cong \begin{cases} \tau_\nu B_\nu(T) & \text{if } \tau_\nu \ll 1 \\ B_\nu(T) & \text{if } \tau_\nu \gg 1 \end{cases}$$

where $\tau_\nu(z)$ is the → optical depth (see section 10.7.1.) of the plasma sheath of thickness z.

References: [27], [48], [54], [64]

10.3. RADIATION TEMPERATURE

In the absence of TE, it is convenient to express the the radiation intensity in terms of the ⟶ Planck function through the relation

$$L_\nu(\mathbf{r},\Omega) = B_\nu(T_{rad}) = \frac{2\,h\,\nu^3}{c^2}\,\frac{1}{\exp\left(\frac{h\nu}{kT_{rad}}\right) - 1}$$

In general, $T_{rad} = T_{rad}(\nu,\mathbf{r},\Omega)$. T_{rad} is called radiation temperature. By analogy with the definition of the mean intensity $\overline{L}_\nu$, a mean radiation temperature, $\overline{T}_{rad}$, can be introduced by

$$\overline{L}_\nu(\mathbf{r}) = B\,(\overline{T}_{rad}) \qquad \text{or} \qquad \rho_\nu(\mathbf{r}) = u_\nu(\overline{T}_{rad})$$

where $T_{rad} = T_{rad}(\nu,\mathbf{r})$.

Reference: [48]

10.4. EMISSION

10.4.1. CONTINUUM SPECTRA

10.4.1.1. ELECTRON-ION FREE-FREE RADIATION

Under the assumptions

(i) electrons have a → Maxwellian velocity distribution at temperature T_e

(ii) binary collision hypotheses: the radiating electron is in the field of only one ion

the spectral emission coefficient (monochromatic bulk radiance) is given by

$$\varepsilon^{ei}_{\nu,ff} = C_1 \frac{Z^2 n_z n_e}{T_e^{1/2}} \bar{G}_z(\nu,T_e)$$

where

$$C_1 = K_\varepsilon^3 \frac{(2\pi)^{1/2}}{3\ 3^{1/2}} \frac{8\ e^6}{c^3\ m_e^{3/2}\ k^{1/2}} = \begin{cases} 5.44 \times 10^{-52}\ \mathrm{J\ m^3\ K^{1/2}/sr} & (SI) \\ 5.44 \times 10^{-39}\ \mathrm{erg\ cm^3\ K^{1/2}/sr} & (CGS) \end{cases}$$

	$\varepsilon^{ei}_{\nu,ff}$	n_z	n_e	T_e	G_z
SI	$W/(m^3\ Hz\ sr)$	m^3	m^3	K	-
CGS	$erg/(s\ cm^3\ Hz\ sr)$	cm^3	cm^3	K	-

n_z is the density of ions of charge state Z,

$$\bar{G}_z(\nu,T_e) = \int_{x_0}^{\infty} exp(-x)\ G_z(u_0)\ dx$$

is the Gaunt factor in classical theory with

$$x = \frac{m_e v_e}{2\ kT_e}, \quad x_0 = \frac{h\nu}{kT_e}, \quad u_0 = \frac{2\pi\ \nu\ p_0}{v_e}$$

and

$$p_0 = K_\varepsilon \frac{Z\ e^2}{m_e\ v_e}$$

is the impact parameter for a 90° deflecion. v_e is the electron velocity and $G_z(u_0)$ is a function close to unity, calculated by Grant [158].

For high frequencies, $h\nu/kT_e \gg 1$, Kramers' formula holds in which

$$g_z(u_0) \cong 1 \quad \text{and} \quad \overline{G}_z(\nu,T_e) = exp\left(-\frac{h\nu}{kT_e}\right)$$

For low and moderate frequencies, $\hbar\omega_{pl} < h\nu < kT_e$, the Gaunt factor can be expressed in the form [31], [43]

$$\overline{G}_z(\nu,T_e) = \frac{\sqrt{3}}{\pi}\left[\ln\left(K_\varepsilon^{-1} \frac{(2\ kT_e)^{3/2}}{Z\ e^2\ \pi\ \nu\ m_e^{1/2}}\right) - \frac{5}{2}\gamma^{EM}\right]$$

$$= \frac{\sqrt{3}}{\pi}\left[\ln\left(2.098 \times 10^8 \frac{T_e^{3/2}}{Z\ \nu}\right) - \frac{5}{2}\gamma^{EM}\right]$$

with $\gamma^{EM} = -\int_0^\infty e^{-x}\, lg\, x\, dx = 0.57722$, the Euler-Mascheroni constant

	$\overline{G}_z$	T_e	ν
SI	-	K	s^{-1}
CGS	-	K	s^{-1}

For very low frequencies, $h\nu/kT_e \rightarrow 0$, $\overline{G}_z(\nu,T_e)$ becomes much greater than unity. The spectrum is independent of frequency, except for a logarithmic dependence on $\overline{G}$, up to $\nu \cong kT_e/h$. For higher frequencies, $\varepsilon_{\nu,ff}^{ei}$ falls to zero.

The Gaunt factor was originally introduced by Kramers [49] to take account for discrepancies between the classical and quantum mechanical treatment of electron bremsstrahlung. He obtained

$$G_z(\nu,v_e) = \frac{\pi\sqrt{3}}{4}\, i\,\Omega\, \mathfrak{H}_{i\Omega}^{(1)}(i\Omega) \left(\frac{d\mathfrak{H}_{i\Omega}^{(1)}(u)}{du}\right)_{u=i\Omega} \xrightarrow[\nu \ll v_e/b_{90}]{} \frac{\sqrt{3}}{\pi} \ln\left(\frac{4\pi\ \ v_e}{\gamma\ b_{90}\ \nu}\right)$$

where

$\mathfrak{H}_{i\Omega}^{(1)}$: first Hankel function of imaginary order and argument

$$\Omega = \frac{\nu}{2\pi}\frac{b_{90}}{v_e}$$

$$b_{90} = K_\varepsilon \frac{Z\ e^2}{m_e\ v_e^2} \qquad : \rightarrow \text{impact paramter for a } 90^\circ \text{ deflection}$$

$$\gamma = exp(\gamma^{EM}) = 1.781... \qquad : \text{Euler's constant}$$

Assuming a → Maxwellian distribution, integration over velocity yields [43]

$$\overline{G}_z(\nu,T_e) \xrightarrow[\nu \ll v_{th}/\overline{b}_{90}]{} \frac{\sqrt{3}}{\pi}\, \ln\left(\frac{1}{3}\left(\frac{2}{\gamma}\right)^{5/2} \frac{v_{th}}{\overline{b}_{90}\,\nu}\right)$$

$$= \frac{\sqrt{3}}{\pi}\, \ln\left(\frac{1}{3}\left(\frac{2}{\gamma}\right)^{5/2} \frac{\omega_{pl}}{\omega}\,\frac{\lambda_D}{\overline{b}_{90}}\right)$$

where

$v_{th} = \left(kT_e/m_e\right)^{1/2}$ → thermal velocity

$\overline{b}_{90} = K_\varepsilon \dfrac{Z e^2}{3\, kT_e}$ → velocity-averaged impact parameter for a 90° deflection

ω_{pl} → plasma frequency

λ_D → Debye length

In his quantum mechanical approach, Gaunt [174] obtained

$$G_z(\nu,v_e) \xrightarrow[\hbar\omega \ll m_e v_e^2/2]{} \frac{\sqrt{3}}{\pi}\, \ln\left(\frac{4\left(\frac{1}{2} m_e v_e^2\right)}{\hbar\,\omega}\right)$$

$$\overline{G}_z(\nu,T_e) \xrightarrow[\hbar\omega \ll kT_e]{} \frac{\sqrt{3}}{\pi}\, \ln\left(\frac{4\, kT_e}{\gamma\, \hbar\omega}\right)$$

The Gaunt factor, $\overline{G}$, and the Coulomb logarithm, $\ln \Lambda$, are equivalent factors arising in the bremsstrahlung and conductivity theories. They are related by ([42] p.253)

$$\overline{G} \equiv \frac{\sqrt{3}}{\pi}\, \ln \Lambda$$

Formulae for the Gaunt factor in the limiting cases of low (high) T_e and n_e can be obtained from the table of asymptotic expressions for the → Coulomb logarithm (see section 3.5.2.). The expressions should be multiplied by

$$\exp\left(-\, \hbar\omega/kT_e\right)$$

resulting from the fact that only those frequencies for which $\hbar\omega \le m_e\, v_e^2 \sim kT_e$ will be emitted. It is customary to write this factor explicitly since the remaining $\overline{G}$ is rather insensitive to frequency. Some authors, however, include the exponential factor, termed as the Cillie exponential factor, in the Gaunt factor. It is negligible in the Rayleigh-Jeans limit

$$\hbar\omega \ll kT_e \qquad \text{or} \qquad \nu \ll 2.42 \times 10^{14}\ T_e$$

where ν is in Hz and T_e in eV.

References: [2], [27], [31], [42], [43], [52], [174]

10.4.1.2. ELECTRON-ATOM FREE-FREE RADIATION

Assuming billiard ball type electron-atom collisions, the spectral emission coefficient can be approximated by

$$\varepsilon^{ea}_{\nu,ff} = C_2\ n_a\ n_e\ T_e^{3/2}\ \overline{G}(\nu,T_e)$$

where

$$C_2 = K_\varepsilon\ \frac{32\ e^2}{3\ c^3}\left(\frac{k}{2\pi\ m_e}\right)^{3/2} = \begin{cases} 3.42 \times 10^{-43}\ \text{J m} / (\text{K}^{3/2}\ \text{sr}) & (SI) \\ 3.42 \times 10^{-34}\ \text{erg cm} / (\text{K}^{3/2}\ \text{sr}) & (CGS) \end{cases}$$

	$\varepsilon^{ea}_{\nu,ff}$	n_a	n_e	T_e	$\overline{G}$
SI	W/(m^3 Hz sr)	m^3	m^3	K	m^2
CGS	erg/(s cm^3 Hz sr)	cm^3	cm^3	K	cm^2

n_a is the density of neutral atoms

$$\overline{G}(\nu,T_e) = \int_{x_0}^{\infty} exp(-x)\ x^2\ \sigma(v_e)\ dx$$ is the Gaunt factor

with $x = \dfrac{m_e\ v_e}{2\ kT_e}$, $x_0 = \dfrac{h\ \nu}{kT_e}$, $\sigma(v_e)$ is the electron-atom impact cross section, and v_e is the electron velocity. The Gaunt factor can be simplified by choosing a constant average value for the cross section:

$$\overline{G}(\nu,T_e) = \overline{\sigma}(T_e)\left[1 + \left(1 + \frac{h\nu}{kT_e}\right)^2\right] exp\left(-\frac{h\nu}{kT_e}\right)$$

For <u>high frequencies</u>, $h\nu/kT_e \gg 1$,

$$\varepsilon^{ea}_{\nu,ff} = C_2\, n_a\, n_e\, T_e^{3/2}\, \bar{\sigma}(T_e) \left(1 + \frac{h\nu}{kT_e}\right)^2 \exp\left(-\frac{h\nu}{kT_e}\right)$$

For <u>low frequencies</u>, $h\nu/kT_e \ll 1$,

$$\varepsilon^{ea}_{\nu,ff} = C_2\, n_a\, n_e\, T_e^{3/2}\, \bar{\sigma}(T_e)$$

In the case of dense and weakly ionized plasmas, electron-atom free-free radiation can become more important than electron-ion free-free radiation.
<u>References</u>: [2], [52]

10.4.1.3. RECOMBINATION RADIATION (FREE-BOUND RADIATION)

The formation of an ion of charge state (Z-1) and the excited state *m* by the capture of a free electron with velocity v_e is accompanied by the emission of a photon, $h\nu$, satisfying

$$h\nu = \frac{1}{2}\, m_e\, v_e^2 + \left(E_{z-1,\infty} - E_{z-1,m}\right)$$

If the electron distribution is assumed to be continuous (e.g. Maxwellian), then the spectrum of photons will be continuous down to the frequency limit

$$\nu_m = \frac{1}{h}\left(E_{z-1,\infty} - E_{z-1,m}\right)$$

The recombination emission coefficient is obtained by summing the spectral emission coeffients of all the levels *m* down to a certain level m_0. For a → Maxwellian electron distribution, this gives

$$\varepsilon_{\nu,fb} = C_3\, \frac{n_e\, n_{z,1}}{T_e^{3/2}}\, \nu^3 \exp\left(-\frac{h\nu}{kT_e}\right) \sum_{m \geq m_0} \frac{g_{z-1,m}}{g_{z,1}}\, Q_{z-1,m}(\nu)$$

where

$$C_3 = \frac{h^4}{c^2 \left(2\pi\, m_e\, k\right)^{3/2}} = \begin{cases} 3.05 \times 10^{-72}\ \mathrm{J\ m\ s^3\ K^{3/2}} & (SI) \\ 3.05 \times 10^{-63}\ \mathrm{erg\ cm\ s^3\ K^{3/2}} & (CGS) \end{cases}$$

	$\varepsilon_{\nu,ff}$	$n_{z,1}$	n_e	T_e	ν	$Q_{z,m}$	$g_{z,m}$
SI	$W/(m^3\ Hz\ sr)$	m^3	m^3	K	Hz	m^2	-
CGS	$erg/(s\ cm^3\ Hz\ sr)$	cm^3	cm^3	K	Hz	cm^2	-

$n_{z,m}$ and $g_{z,m}$ are respectively the density and statistical weight of particles in charge state Z and excited state *m* and $Q_{z-1,m}(\nu)$ denotes the cross-section for photo-ionization of a (Z-1) charged ion in the excited state *m*.
By using specific expressions for the photo-ionization cross-section, the recombination emission coefficient can written more explicitly:

(A) Hydrogenic atoms and ions

$$\varepsilon_{\nu,fb} = C_4 \frac{n_e\, n_z}{T_e^{3/2}} Z^4 \exp\left(-\frac{h\nu}{kT_e}\right) \sum_{m \geq m_0} G_m(\nu)\, m^{-3} \exp\left(\frac{Z^2\, E_{H,\infty}}{m^2\, kT_e}\right)$$

with

$$C_4 = K_\varepsilon^5 \frac{32\,(2\,\pi^5)^{1/2}\, m_e\, e^{10}}{c^3 \left(3\, m_e\, k\right)^{3/2} h^2} = \begin{cases} 1.72 \times 10^{-46}\ J\ m^3\ K^{3/2} & (SI) \\ 1.72 \times 10^{-33}\ erg\ cm^3\ K^{3/2} & (CGS) \end{cases}$$

m is the principal quantum number, m_0 is determined by

$$m_0 = Z \left(\frac{E_{H,\infty}}{h\nu}\right)^{1/2}$$

where $E_{H,\infty}$= 13.60 eV is the ionization energy of atomic hydrogen; $G_m(\nu)$ is the Gaunt factor, which is very near unity in the visible region for $m \geq 3$. $G_m(\nu)$ is tabulated e.g. in [159]. Due to the hydrogenic level structure, $n_{z,1}$ has been replaced simply by n_z in the above formula.
Atomic levels are very close to one another for $m \geq m_0$= 4 ; at low frequencies,

$$\nu < \frac{Z^2\, E_{H,\infty}}{h\, m_0^2}$$

(which corresponds to λ > 1.4 μm), the recombination emission coefficient can be simplified as follows,

$$\varepsilon_{\nu,fb} = C_1 \frac{Z^2 n_z n_e}{T_e^{1/2}} \left[1 - \exp\left(- \frac{h\nu}{kT_e}\right)\right]$$

Here, the Gaunt factor is set to unity and the constant C_1 can be found in section 10.4.1.1.

(B) Non-hydrogenic atoms and ions

$$\varepsilon_{\nu,fb} = C_1 \frac{Z^2 n_z n_e}{T_e^{1/2}} \left[1 - \exp\left(- \frac{h\nu}{kT_e}\right)\right] \frac{g_{z,1}}{\sigma_z(T_e)} \xi_z(\nu,T_e)$$

Because of the non-hydrogenic level structure, $n_{z,1}$ is expressed by $n_{z,1} = n_z \frac{g_{z,1}}{\sigma_z(T_e)}$, with σ_z being the → partition function of the Z-fold ionized atom. At low temperatures, however, it is always true that $\sigma_z \cong g_z$.

The Biberman factor, $\xi_z(\nu,T_e)$ (also denoted as the Biberman-Gaunt factor or the Biberman-Schlüter factor) takes the electronic structure of an atom (in charge state Z) into consideration. It's value is near unity and depends on frequency but weakly on temperature.

References: [2]-[4], [52], [160]

10.4.1.4. TOTAL FREE-FREE AND RECOMBINATION CONTINUUM

Neglecting free-free electron-atom radiation, the total spectral emission coefficient can be written as

$$\varepsilon_{\nu,total} = \varepsilon^{ei}_{\nu,ff} + \varepsilon_{\nu,fb} = C_1 \frac{Z^2 n_z n_e}{T_e^{1/2}} \left[G_z(\nu,T_e) + \xi_z(\nu,T_e)\left[1 - \exp\left(- \frac{h\nu}{kT_e}\right)\right]\right]$$

At high frequencies, $h\nu/kT_e \gg 1$, $G_z(\nu,T_e) = \exp\left(- \frac{h\nu}{kT_e}\right)$, the continuum is determined by recombination radiation

$$\varepsilon_{\nu,total} = C_1 \frac{Z^2 n_z n_e}{T_e^{1/2}} \xi_z(\nu,T_e)$$

At low frequencies, $h\nu/kT_e \ll 1$, the continuum is determined by free-free radiation,

$$\varepsilon_{\nu,\mathrm{total}} = C_1 \frac{Z^2 n_z n_e}{T_e^{1/2}} \bar{G}_z(\nu,T_e)$$

Following Biberman's approach ([3] and [33]), atomic levels can be classified into two groups:

(i) levels which differ significantly from the hydrogenic structure (significant quantum defect): the Burgess and Seaton theory [20] or an alternative quantum mechanical theory should be applied,

(ii) nearly hydrogenic levels (weak quantum defect): the Kramers-Unsöld theory can be applied [28], [49]-[51]. In the latter, close adjacent levels below the ionization limit are treated as a quasi-continuum down to the excited state m_0

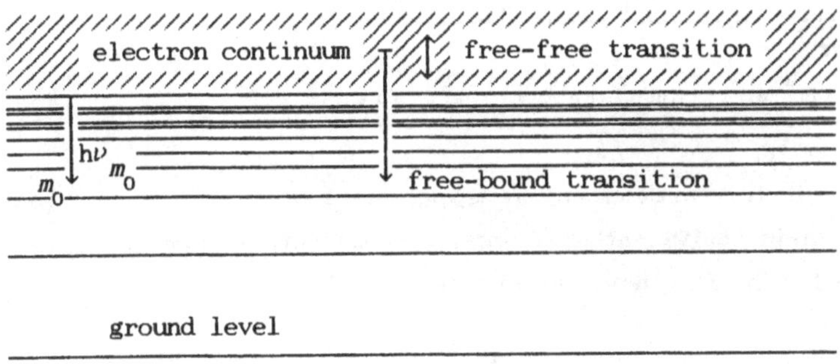

Then the spectral emission coefficient for free-free and recombination radiation is given by

$$\varepsilon_{\nu,\mathrm{total}} = C_1 \frac{Z^2 n_z n_e}{T_e^{1/2}} \xi_z(\nu,T_e) \qquad \text{for } \nu < \nu_{m_0}$$

$$\varepsilon_{\nu,\mathrm{total}} = C_1 \frac{Z^2 n_z n_e}{T_e^{1/2}} \xi_z(\dot{\nu},T_e) \exp\left(-\frac{h\left(\nu - \nu_{m_0}\right)}{kT_e}\right) \qquad \text{for } \nu_{m_0} < \nu < \nu_l$$

The contribution of isolated levels, l, below m_0, $\nu > \nu_{m_0}$, to the spectral emission coefficient must be added separately. The frequency ν_{m_0} is related to the level energy by

$$\nu_{m_0} = \frac{1}{h}\left(E_{z-1,\infty} - E_{z-1,m_0}\right)$$

In plasmas, the reduction of the ionization energy, $E_{z-1,\infty} - \Delta E$, must be taken into account. The limiting frequency ν_{m_0} must be replaced by the corrected value, $\nu_{m_0} - \Delta E/h$.

References: [2], [3], [27], [28], [33], [39], [49]-[53]

10.4.2. LINE RADIATION (BOUND-BOUND TRANSITIONS)

The total emission coefficient (also: bulk radiance) of a spectral line is given by

$$\varepsilon_{ul} = \frac{1}{4\pi} n_u A_{ul} h\nu$$

	ε_{ul}	n_u	A_{ul}	$h\nu$
SI	W/m^3 sr	m^{-3}	s^{-1}	J
CGS	erg/s cm^3 sr	cm^{-3}	s^{-1}	erg

where

n_u : density of particles in the upper state u

A_{ul} : Einstein coefficient of spontaneous transitions $u \longrightarrow l$

$\nu = |E_u - E_l|/h$ frequency at line centre

In case of an equilibrium plasma using the $\rightarrow$ Boltzmann distribution of excited states, we obtain,

$$\varepsilon_{ul} = \frac{1}{4\pi} \frac{g_{z,u}}{\sigma_z} n_z A_{ul} h\nu \exp\left(-\frac{E_{z,u}}{kT}\right)$$

where

$g_{z,u}$: statistical weight of the level u for a particle in ionization state Z

σ_z : partition function

n_z : (total) number density of particles with charge Z

$E_{z,u}$: excitation energy of the excited level u

ε_{ul} is the integral over the frequency area of the spectral line

$$\varepsilon_{ul} = \int_{(\nu)} \varepsilon_\nu \, d\nu \qquad \text{with} \qquad \varepsilon_\nu = \frac{1}{4\pi} n_u A_{ul} h\nu P(\nu)$$

where $P(\nu)$ is the $\longrightarrow$ spectral line shape (see 10.5.).

Conversion between transition probability A_{ul} , oscillator strength f_{lu} , and line strength, S_{lu}

Notation:

$\mathfrak{E}_d$: electric dipole (allowed) transitions

$\mathfrak{E}_q$: electric quadropole (forbidden) transitions

$\mathfrak{M}_d$: magnetic dipole (forbidden) transitions

λ : wavelength

$g_{u/l}$: statistical weight of the upper and lower states

The relation between the transition probability, A_{ul}, and the oscillator strength f_{lu} is given by the Ladenburg formula (allowed electric dipole transitions)

$$A_{ul} = K_\varepsilon \frac{8 \pi^2 e^2 \nu^2}{m_e c^3} \frac{g_l}{g_u} f_{lu} = K_\varepsilon \frac{8 \pi^2 e^2}{m_e c \lambda^2} \frac{g_l}{g_u} f_{lu}$$

From time to time, the line strength S_{lu} is used in place of the transition probability. The relation is

$$S_{lu} = \frac{3 h \lambda g_u}{64 \pi^4} A_{ul} = K_\varepsilon \frac{3 h e^2 \lambda g_l}{8 \pi^2 m_e c} f_{lu}$$

Numerical values are given in the table (next page). The units are

	A_{ul}	f_{lu}	S_{lu}	λ	g
SI	s^{-1}	-	atomic units	nm	-
CGS	s^{-1}	-	atomic units	nm	-

atomic units for S_{lu} : $e^2 a_0^2$, where e is the elementary charge and a_0 is the → Bohr radius.

The line strength is symmetrical with respect to $u \longleftrightarrow l$: $S_{ul} = S_{lu}$.
The *emission* oscillator strength, f_{ul}, is related to the *absorption* oscillator strength, f_{lu}, by

$$f_{ul} = - \frac{g_l}{g_u} f_{lu}$$

	transition probability A_{ul}	oscillator strength f_{lu}	line strength S_{lu}
$A_{ul} =$	$1\ A_{ul}$	$\frac{6.670 \times 10^{13}}{\lambda^2} \frac{g_l}{g_u} f_{lu}$	$\mathfrak{E}_d$: $\frac{2.026 \times 10^{15}}{g_u\ \lambda^2} S_{lu}$ $\mathfrak{E}_q$: $\frac{1.679 \times 10^{13}}{g_u\ \lambda^5} S_{lu}$ $\mathfrak{M}_d$: $\frac{2.697 \times 10^{10}}{g_u\ \lambda^3} S_{lu}$
$f_{lu} =$	$1.4992 \times 10^{-14}\ \lambda^2 \frac{g_u}{g_l} A_{ul}$	$1\ f_{lu}$	$\mathfrak{E}_d$: $\frac{30.37}{g_l\ \lambda} S_{lu}$ $\mathfrak{E}_q$: $\frac{0.2519}{g_l\ \lambda^3} S_{lu}$ $\mathfrak{M}_d$: $\frac{4.043 \times 10^{-4}}{g_l\ \lambda} S_{lu}$
$S_{lu} =$	$\mathfrak{E}_d$: $4.935 \times 10^{-16}\ g_u\ \lambda^3\ A_{ul}$ $\mathfrak{E}_q$: $5.953 \times 10^{-13}\ g_u\ \lambda^5\ A_{ul}$ $\mathfrak{M}_d$: $3.707 \times 10^{-11}\ g_u\ \lambda^3\ A_{ul}$	$\mathfrak{E}_q$: $3.292 \times 10^{-2}\ g_l\ \lambda\ f_{lu}$ $\mathfrak{E}_q$: $3.971\ g_l\ \lambda^3\ f_{lu}$ $\mathfrak{M}_d$: $2472\ g_l\ \lambda\ f_{lu}$	$1\ S_{lu}$

The <u>lifetime of an excited level</u> *u* is given by

$$\tau_u = \frac{1}{\sum\limits_{l<u} A_{ul}}$$

The relation between the Einstein coefficient of spontaneous emission to that of induced emission is

$$\frac{A_{ul}}{B_{ul}} = \frac{8\pi\, h\, \nu^3}{c^2}$$

and the Einstein coefficient for (induced) absorption, B_{lu}, is related to B_{ul} and f_{lu} by

$$B_{lu} = \frac{g_u}{g_l} B_{ul} = K_\varepsilon \frac{e^2\, \pi}{m_e\, c\, h\, \nu} f_{lu} = \begin{cases} 4.005 \times 10^{27}\, f_{lu}/\nu & (SI) \\ 4.005 \times 10^{24}\, f_{lu}/\nu & (CGS) \end{cases}$$

	A_{ul}	B_{ul}	B_{lu}	f_{lu}	f_{ul}	$g_{u/l}$	ν
SI	s^{-1}	$(J/m^2\ Hz)^{-1}$	$(J/m^2\ Hz)^{-1}$	-	-	-	s^{-1}
CGS	s^{-1}	$(erg/cm^2\ Hz)^{-1}$	$(erg/cm^2\ Hz)^{-1}$	-	-	-	s^{-1}

Kramer's quasi-classical formula for hydrogen-like atoms:

$$f_{lu} \cong \frac{32}{3\pi\ \ 3^{1/2}} \frac{1}{l^5\, u^3 \left(l^{-2} - u^{-2}\right)^3}$$

where

u : principal quantum number of the upper state

l : principal quantum number of the lower state

Rydberg formula for hydrogen:

$$h\nu = Ry \left(\frac{1}{l^2} - \frac{1}{u^2}\right)$$

where

$$Ry = \frac{Ry_\infty}{1 + \frac{m_e}{m_a}}$$

with

$$Ry_\infty = K_\varepsilon^2 \frac{2\, \pi^2\, e^4\, m_e}{h^2} = 13.60\ eV = \begin{cases} 2.180 \times 10^{-18}\ J & (SI) \\ 2.180 \times 10^{-11}\ erg & (CGS) \end{cases}$$

$$\hat{=} \begin{cases} 1.097373 \times 10^{7}\ m^{-1} & (SI) \\ 1.097373 \times 10^{5}\ cm^{-1} & (CGS) \end{cases}$$

and m_a denotes the mass of the hydrogen-like particle.

Wigner-Kirkwood rule for 1 electron jump ([78] p.56)

$$\sum f = -\frac{1}{3}\,\frac{\ell\,(2\ell-1)}{2\ell+1} \qquad \text{for } \ell \longrightarrow \ell - 1$$

$$\sum f = -\frac{1}{3}\,\frac{(\ell+1)\,(2\ell+3)}{2\ell+1} \qquad \text{for } \ell \longrightarrow \ell - 1$$

where ℓ denotes the orbital quantum number, for example

$$\sum f = -2/5 \qquad \text{for } d \longrightarrow np$$

$$\sum f = 10/9 \qquad \text{for } p \longrightarrow nd$$

This rule applies precisely for hydrogen. For complicated spectra it may be used as an approximative formula.

References: [27], [28] p.336, [32], [52] p.162, [78], [161]

10.5. SPECTRAL LINE SHAPES

All profiles are given in the ν-notation. To transform to the λ-notation simply use $\nu_0 = c/\lambda_0$ and $|\Delta\nu| = |\Delta\lambda|\ c/\lambda_0^2$.

10.5.1. DOPPLER PROFILE (GAUSS PROFILE)

$$P_D(\nu) \sim exp\left[-\left(\frac{\Delta\nu}{\Delta\nu_D}\right)^2\right]$$

$\Delta\nu = \nu - \nu_0$ is the distance from line centre, ν_0 is the frequency at line centre and $\Delta\nu_D$ the half-width at the 1/e value. If P_D is required in terms of the full width at half maximum (FWHM),

$$P_D(\Delta\nu) \sim exp\left[-\left(\frac{2\,(\ln 2)^{1/2}\,\Delta\nu}{\Delta\nu_D^{FWHM}}\right)^2\right]$$

where $\Delta\nu_D^{FWHM} = 2\,(\ln 2)^{1/2}\,\Delta\nu_D$. With the normalization $\int_{-\infty}^{\infty} P_D\, d\nu = 1$,

$$P_D(\Delta\nu) = \frac{1}{\pi^{1/2}\,\Delta\nu_D}\, exp\left[-\left(\frac{\Delta\nu}{\Delta\nu_D}\right)^2\right]$$

or

$$P_D(\Delta\nu) = \frac{2\,(\ln 2)^{1/2}}{\pi^{1/2}\,\Delta\nu_D^{FWHM}}\, exp\left[-\left(\frac{2\,(\ln 2)\,\Delta\nu}{\Delta\nu_D^{FWHM}}\right)^2\right]$$

The Doppler profile is caused by the thermal motion of the emitting (or absor bing) particles.

10.5.2. LORENTZ PROFILE (DISPERSION PROFILE)

$$P_L \sim \frac{1}{1 + \left(\frac{\Delta\nu}{\Delta\nu_L}\right)^2}$$

or

$$P_L \sim \frac{1}{1 + \left[\frac{2\ \Delta\nu}{\Delta\nu_L^{FWHM}}\right]^2}$$

with

$\Delta\nu = \nu - \nu_0$: distance from line centre

$\Delta\nu_L$: (half) width at half maximum of a Lorentz-profile

$\Delta\nu_L^{FWHM} = 2\ \Delta\nu_L$: full width at half maximum

With the normalization

$$\int_{-\infty}^{\infty} P_L(\nu-\nu_0)\ d\nu = 1$$

we obtain

$$P_L(\Delta\nu) = \frac{1}{\pi} \frac{\Delta\nu_L}{(\Delta\nu)^2 + (\Delta\nu_L)^2}$$

or

$$P_L(\Delta\nu) = \frac{1}{2\pi} \frac{\Delta\nu_L^{FWHM}}{(\Delta\nu)^2 + (\Delta\nu_L^{FWHM}/2)^2}$$

The Lorentz profile is caused by

- natural broadening
- collisional broadening; in particular:
  - Stark broadening: (collisions with electrons and ions)
  - resonance broadening (Holtzmark broadening): collisions with other absorbing or emitting atoms of the same kind)
  - Van der Waals broadening (Lorentz broadening): collisions with atoms of a different kind)

Remark:

There are several theoretical approaches to spectral line broadening. The impact approximation and the quasi-static approximation are two asymptotic cases of quasi-classical theory.

The impact approximation is applicable in low-density plasmas where the radiator is assumed to be most the time without any disturbance and during the short collision time it is interacting with only one colliding particle.

The quasi-static approximation is applicable for plasmas of moderate density

where the radiator is under strong, slowly varying disturbances.
Impact and quasi-static approximation yield different line profiles for an van der Waals interaction potential: the former yields an → Lorentz profile and the latter an asymmetric profil of the form

$$P_w(\Delta\nu) = \begin{cases} -\left(\dfrac{\Delta\nu_0}{(\Delta\nu)^3}\right)^{1/2} exp\left(-\dfrac{\pi}{4}\dfrac{\Delta\nu_0}{\Delta\nu}\right) & \text{for } \nu < \nu_0 \\ 0 \quad \text{for } \nu \geq \nu_0 \end{cases}$$

Under these conditions, experimental line profiles are usually described by a convolution integral of the kind [190]

$$P(\Delta\nu) = \int_{-\infty}^{\infty} P_L(\Delta\nu - \xi)\, P_w(\Delta\nu)\, d\xi$$

References: [56] chapter 4, [57], [121], [195]

10.5.3. VOIGT PROFILE

The Voigt profile is defined as the convolution of a → Doppler profile with a → Lorentz profile

$$P_V(a,x) = -\int_{-\infty}^{\infty} \frac{exp(-y^2)}{a^2 + (x+y)^2}\, dy$$

where a is the "damping ratio", representing the ratio of the width of a Lorentz profile to a Doppler profile

$$a = \frac{\Delta\nu_L}{\Delta\nu_D} = \frac{\Delta\nu_L^{FWHM}}{\Delta\nu_D^{FWHM}}\,(\ln 2)^{1/2}$$

x is the scaled frequency:

$$x = \frac{\Delta\nu}{\Delta\nu_D} = \frac{\Delta\nu}{\Delta\nu_D^{FWHM}}\, 2\, (\ln 2)^{1/2}$$

For $a \rightarrow 0$ one obtains the Doppler profile, for $a \rightarrow \infty$ the Lorentz profile is obtained. If $0 < a < 1$, the Voigt profile consists of a core governed by the Doppler profile and wings described by a Lorentz profile.

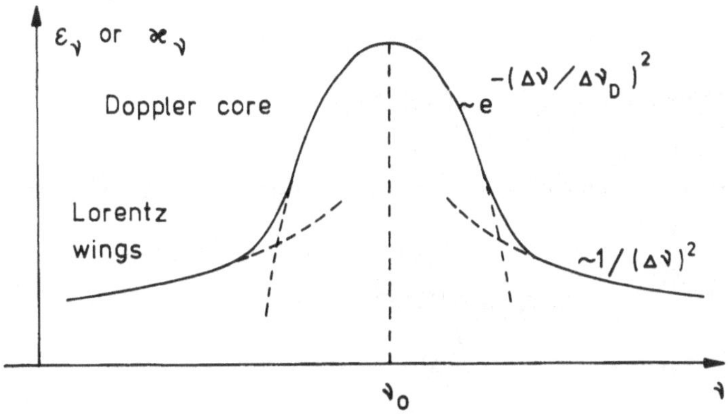

References: [39], [52], [54], [55]

10.6. SPECTRAL LINE WIDTH

With reference to the figure below, the half width at 1/e should be distinguished from the full width at half maximum (FWHM).

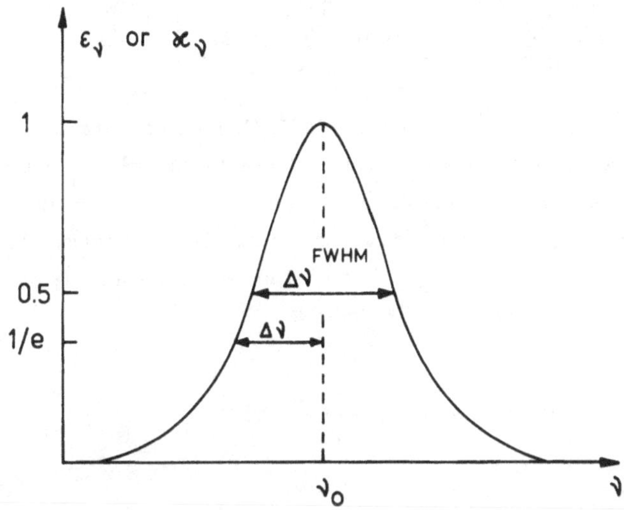

10.6.1. NATURAL WIDTH

The full width at half maximum is given by

$$\Delta\nu_{nat}^{FWHM} = \frac{1}{2\pi}\left(\frac{1}{\tau_u} + \frac{1}{\tau_l}\right) = \frac{\gamma}{2\pi}$$

where τ_u : lifetime of the upper state
τ_l : lifetime of the lower state
γ : damping constant

In the case of a resonance line, $l = 1$ and $\tau_1 = \infty$. Hence, $\Delta\nu_{nat}^{FWHM} = \frac{1}{2\pi} A_{u1}$.

References: [52] p.161, [55] p.100

10.6.2. DOPPLER WIDTH

$$\Delta\nu_D^{FWHM} = 2\,(\ln 2)^{1/2}\ \frac{\nu_0}{c}\left(\frac{2\ kT}{m}\right)^{1/2}$$

where T is the temperature of absorbing or emitting particles and m denotes their mass. The half width at 1/e is

$$\Delta\nu_D = \frac{\nu_0}{c}\left(\frac{2\ kT}{m}\right)^{1/2}$$

References: [39], [55]

10.6.3. STARK WIDTH

According to Griem's theory, Stark broadening will result in a full width at half maximum of a Lorentz profile given by

$$\Delta\nu_{Stark}^{FWHM} = \left[1 + 1.75\ \alpha\left(\frac{n_e}{10^{22}}\right)^{1/4}\left(1 - 0.75\ \frac{\bar{l}_i}{\lambda_{D,e}}\right)\right] 2\ \Delta\nu_e\left(\frac{n_e}{10^{22}}\right)$$

in *SI*. In *CGS* the value 10^{22} should be replaced by 10^{16}. The parameters α and $\Delta\nu_e$ are weak functions of the electron temperature and are tabulated by Griem ([57] appendix IVa). The broadening by electrons is given by the term containing $\Delta\nu_e$ with the term in α giving the ion contribution. In this term, the ratio of the mean ion distance $\bar{l}_i = (3/4\pi\ n_i)^{1/3}$ and the → electronic Debye length, $\lambda_{D,e}$, is important. The formula given is valid if

$$\alpha\left(\frac{n_e}{10^{22}}\right)^{1/4} \le 0.5 \qquad \text{and} \qquad \frac{\bar{l}_i}{\lambda_{D,e}} \le 0.8$$

References: [56], [57]

10.6.4. RESONANCE WIDTH

$$\Delta\nu_{resonance}^{FWHM} = C\ \frac{f_{lu}\ n_l}{\nu_0}\left(\frac{g_l}{g_u}\right)^{1/2}$$

with

$$C = K_\varepsilon\ \frac{3\ e^2}{\pi\ m_e} = \begin{cases} 2.418 \times 10^2\ m^3/s^2 & (SI) \\ 2.418 \times 10^8\ cm^3/s^2 & (CGS) \end{cases}$$

	$\not\alpha\nu$	ν_0	n_l	$g_{l/u}$	f_{lu}
SI	s^{-1}	s^{-1}	m^{-3}	-	-
CGS	s^{-1}	s^{-1}	cm^{-3}	-	-

References: [39], [52], [55], [56], [57]

10.7. ABSORPTION

10.7.1. GENERAL

The intensity, I_ν, of incident radiation of frequency ν will be reduced by dI in an absorbing layer of thickness ds according to the absorption law

$$\frac{dI_\nu}{I_\nu} = - \varkappa_\nu \, ds$$

where $\varkappa_\nu$ is the → absorption coefficient at a given frequency ν.

	I_ν	$\varkappa_\nu$	ds	τ_ν
SI	J/m^2 sr	m^{-1}	m	-
CGS	erg/cm^2 sr	cm^{-1}	cm	-

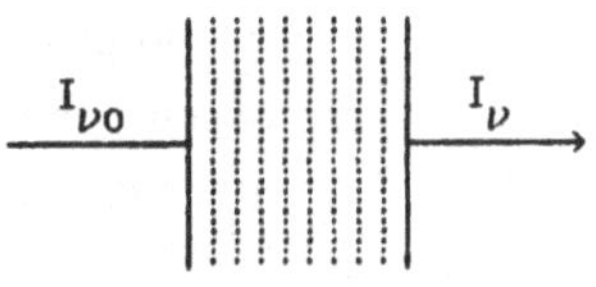

Integration gives

$$I_\nu = I_{\nu 0} \exp\left(- \tau_\nu\right)$$

Here τ_ν is the optical depth of the absorbing layer with geometrical thickness s

$$\tau_\nu = \int_0^s \varkappa_\nu \, ds$$

More specific defintions may also be used:

optical depth of a spectral line $\quad \tau_\nu = \int_{-\infty}^{s} \varkappa_\nu \, ds$

optical depth of the continuum $\quad \tau_{con} = \int_{-\infty}^{s} \varkappa_{con} \, ds$

total optical depth $\quad \tau = \int_{-\infty}^{s} (\varkappa_\nu + \varkappa_{con}) \, ds$

where

$\varkappa_\nu$: line absorption coefficient (depends strongly on the distance, $\Delta\nu = \nu - \nu_0$ from the line centre at frequency ν_0; $\varkappa_\nu \to 0$ far from line centre)

$\varkappa_{con}$: continuum absorption coefficient (depends weakly on frequency)

According to the → Kirchhoff law of radiation, the absorption coefficient, $\varkappa_\nu$, can be found by using the corresponding → emission coefficient ε_ν

$$\boxed{\varkappa_\nu = \varepsilon_\nu \, B_\nu^{-1}}$$

where B_ν denotes the → Planck function (see section 10.2.).

For a Doppler broadened line, the optical depth at line centre written as

$$\tau = \frac{1}{8\,\pi^{3/2}} \frac{c^3}{\nu_0^3} A_{ul} \frac{g_u}{g_l} n_l \, \ell \left(\frac{m}{2\,kT}\right)^{1/2}$$

where

n_l : density of particles in the lower state

ℓ : characterisitc length of the plasma layer

ν_0 : frequency at line centre ($c^3/\nu_0^3 = \lambda_0^3$)

$(m/2\,kT)^{1/2}$: inverse thermal velocity of absorbing particles

A_{ul} : Einstein coefficient of spontaneous emission

$g_{u/l}$: statistical weight of the upper and lower levels

References: [54], [58]

10.7.2. FREE-FREE TRANSITIONS

Using ε_ν in the quasi-classical approximation (Kramers-Unsöld theory)

$$\varkappa_{\nu,ff} = K_\varepsilon^3 \frac{4\,(2\pi)^{1/2}\, Z^2\, e^6\, n_e\, n_i\, \overline{G}_z(\nu,T_e)}{3^{3/2}\, c\, h\, m_e^{3/2}\, \nu^3\, (kT_e)^{1/2}} \left[1 - \exp\left(-\frac{h\nu}{kT_e}\right)\right]$$

$$= \begin{cases} 3.69 \times 10^{-2}\, Z^2\, n_e\, n_i\, \overline{G}_z(\nu,T_e) \left[\;\right] / \nu^3\, T_e^{1/2} & \textit{(SI)} \\ 3.69 \times 10^{8}\, Z^2\, n_e\, n_i\, \overline{G}_z(\nu,T_e) \left[\;\right] / \nu^3\, T_e^{1/2} & \textit{(CGS)} \end{cases}$$

where $\overline{G}_z$ is the → Gaunt factor for free-free transitions (see 10.4.1.1). Note, the factor $\left[1 - \exp(-h\nu/kT_e)\right]$ takes stimulated emission into account. If $h\nu/kT_e \ll 1$, series expansion gives the approximate formula

$$\varkappa_{\nu,ff} = K_\varepsilon^3 \frac{4\,(2\pi)^{1/2}\,Z^2\,e^6\,n_e\,n_i\,\overline{G}_z(\nu,T_e)}{3^{3/2}\,c\,m_e^{3/2}\,\nu^2\,(kT_e)^{3/2}}$$

$$= \begin{cases} 1.77 \times 10^{-12}\,Z^2\,n_e\,n_i\,\overline{G}_z\,/\,\nu^2\,T_e^{3/2} & (SI) \\ 1.77 \times 10^{-2}\,Z^2\,n_e\,n_i\,\overline{G}_z\,/\,\nu^2\,T_e^{3/2} & (CGS) \end{cases}$$

	$\varkappa_\nu$	T_e	n	ν	Z
SI	m^{-1}	K	m^{-3}	s^{-1}	-
CGS	cm^{-1}	K	cm^{-3}	s^{-1}	-

This is the classical border-line case. Hence, Planck's constant, h, does not occur.

References: [28], [59]

10.7.3. BOUND-BOUND TRANSITIONS

The absorption coefficient for an atomic transition from a lower level, *l*, to an upper level, *u*, is given by

$$\varkappa_\nu(l \rightarrow u) = \sigma_\nu(l \rightarrow u)\; n_l$$

where $\sigma_\nu(l \rightarrow u)$ is the cross-section for a bound-bound absorption and n_l denotes the number density of atomic particles in the lower state.

The absorption coefficient of an absorption line having a Doppler profile can be expressed as

$$\varkappa_{\nu,D} = \varkappa^* \; P_D(\Delta\nu)$$

where $P_D(\Delta\nu)$ is the normalized → Doppler profile $\left(\int P_D \, d\nu = 1 \right)$

$$\varkappa^* = \frac{c^2}{\nu_0^2} \frac{A_{ul}}{8\pi} \frac{g_u}{g_l} n_l$$

With the help of the → Ladenburg formula (see section 10.4.2.), $\varkappa^*$ can be rewritten as a function of the oscillator strength, f_{lu}, and the density n_l,

$$\varkappa^* = C^* \, f_{lu} \, n_l$$

where

$$C^* = K_\varepsilon \frac{\pi e^2}{m_e c} = \begin{cases} 2.654 \times 10^{-6}\ m^2/s & (SI) \\ 2.654 \times 10^{-2}\ cm^2/s & (CGS) \end{cases}$$

	$\varkappa_\nu$	$\varkappa^*$	P_D	f_{lu}	n_l
SI	m^{-1}	$(m\ s)^{-1}$	s	-	m^{-3}
CGS	cm^{-1}	$(cm\ s)^{-1}$	s	-	cm^{-3}

With the → Doppler profile at $\nu = \nu_0$ (line centre), the absorption coefficient is given by

$$\varkappa_{\nu,D}(\nu_0) = \frac{c^2\ A_{ul}\ n_l}{8\ \pi^{3/2}\ \nu_0^2\ \Delta\nu_D} \frac{g_u}{g_l} = \frac{(\ln 2)^{1/2}\ c^2\ n_l\ A_{ul}}{4\ \pi^{3/2}\ \nu_0^2\ \Delta\nu_D^{FWHM}} \frac{g_u}{g_l}$$

The absorption coefficient of an absorption line having a Lorentz profile can be expressed as

$$\varkappa_{\nu,L} = \varkappa^*\ P_L(\Delta\nu)$$

where $P_L(\Delta\nu)$ is the normalized → Lorentz profile $\left(\int P_L\ d\nu = 1 \right)$.

At the line centre, $\Delta\nu = 0$,

$$\varkappa_{\nu,L}(\nu_0) = \frac{c^2\ A_{ul}\ n_l}{8\ \pi^2\ \nu_0^2\ \Delta\nu_L} \frac{g_u}{g_l} = \frac{c^2\ A_{ul}\ n_l}{4\ \pi^2\ \nu_0^2\ \Delta\nu_L^{FWHM}} \frac{g_u}{g_l}$$

The absorption coefficient of an absorption line having a Voigt profile can be expressed as

$$\varkappa_{\nu,V} = \varkappa^*\ P_V(a,x)$$

where $P_V(a,x)$ is the → Voigt profile . In line centre, $\nu = \nu_0$ i.e. $x = 0$,

$$\varkappa_{\nu,V}(\nu_0) = \frac{(\ln 2)^{1/2}\ c^2\ A_{ul}\ n_l}{4\ \pi^{3/2}\ \nu_0^2\ \Delta\nu_V^{FWHM}} \frac{g_u}{g_l}$$

If the damping ratio $a < 0.1$, then $\Delta\nu_V^{FWHM} \cong \Delta\nu_D^{FWHM}$ and thus

$$\varkappa_{\nu,V}(\nu_0) \cong \varkappa_{\nu,D}(\nu_0)$$

If the damping ratio $a > 0$, and $(\Delta\nu)^2 = (\nu - \nu_0)^2 \gg \left(\Delta\nu_V^{FWHM}\right)^2$, that is in the wings of the profile, then

$$P_V(x,a) \longrightarrow P_L \cong \frac{\Delta\nu_V^{FWHM}}{2\pi\ (\Delta\nu)^2}$$

and

$$\varkappa_{\nu,V}(\Delta\nu) = \varkappa^* P_L = \frac{c^2\ A_{ul}\ n_l\ g_u}{8\pi\ \ \nu_0^2} \frac{\Delta\nu_V^{FWHM}}{2\pi\ (\Delta\nu)^2}$$

For hydrogenic atomic systems the <u>absorption coefficient at the line centre</u> may be written as

$$\varkappa_{lu} = \frac{2^7\ \pi}{3^{3/2}} \alpha_F^3\ a_0^2 \left(\frac{Ry_E}{h\nu}\right)^3 Z_{eff}^4 \frac{n_l}{l^5}$$

where

α_F : fine-structure constant (1/137)

a_0 : Bohr radius

Ry_E : Rydberg constant in energy units

Z_{eff} : effective atomic charge

l : principal quantum number of the initial (lower) state

The <u>total absorption coefficient of a spectral line</u> is the frequency integral covered by the line,

$$\int_{(\nu)} \varkappa_\nu\ d\nu = K_\varepsilon \frac{\pi\ e^2\ n_n\ f_{lu}}{m_e\ c} = \frac{\lambda_{ul}\ n_l\ A_{ul}}{8\pi} \frac{g_u}{g_l}$$

This shows that whatever physical processes are responsible for the formation of the absorption line, the integral of the absorption coefficient over frequency remains constant when the density n_l is constant.

<u>References</u>: [33], [39], [52]-[55]

10.8. RADIATION TRANSPORT

10.8.1. PLANCK APPROXIMATION

For nearly transparent plasmas, the mean-free-path of a photon, l_ν, is large in comparison with the characteristic length of the plasma extension, l_{pl},

$$l_\nu \cong \frac{1}{\varkappa'_\nu} \gg l_{pl} \qquad \text{with} \qquad \varkappa'_\nu = \varkappa_\nu \left[1 - \exp\left(-\frac{h\nu}{kT}\right)\right]$$

and the optical depth is small:

$$\tau_\nu = \int \varkappa'_\nu \, ds \cong \varkappa'_\nu \, l_{pl} \ll 1$$

(The term in square brackets takes stimulated emission into account.)
In this case the intensity of the radiation field, L_ν, is much smaller than the equilibrium intensity:

$$L_\nu = \int B_\nu \varkappa'_\nu \exp\left(-\tau_\nu\right) ds \cong \int B_\nu \varkappa'_\nu \, ds \cong B_\nu \tau_\nu \ll B_\nu$$

Consequently,

$$\rho_\nu \ll u_\nu$$

where

ρ_ν → spectral (monochromatic) radiation density

u_ν → spectral (monochromatic) radiation denstity in thermodynamic equilibrium (TE)

If scattering is neglected, the radiation transport equation in TE approximation can be written as follows,

$$\frac{1}{c}\frac{\partial L_\nu}{\partial t} + \vec{\Omega}\, \nabla L_\nu = \varkappa'_\nu \left(B_\nu - L_\nu\right)$$

or

$$\frac{\partial \rho_\nu}{\partial t} + \nabla S_\nu = c\, \varkappa'_\nu \left(u_\nu - \rho_\nu\right)$$

where

$$S_\nu = \int_{4\pi} L_\nu \vec{\Omega}\, d\Omega \equiv 2\pi F_\nu$$

is the monochromatic radiation flux vector. In the Planck approximation (transparent plasma), the quasi-stationary part of the radiation transport eq. gives

$$\nabla S_\nu = c\, \varkappa'_\nu \left(u_\nu - \rho_\nu\right) \cong c\, \varkappa'_\nu\, u_\nu$$

Integration over frequency yields

$$\nabla S = c\, \bar{\varkappa}_{planck}\, u$$

where u is the → radiation density in TE (see section 10.2.) and

$$\bar{\varkappa}_{Planck} = \frac{\int_0^\infty \varkappa'_\nu\, u_\nu\, d\nu}{\int_0^\infty u_\nu\, d\nu}$$

is the Planck mean absorption coefficient.

References: [48], [58], [191]-[194]

10.8.2. ROSSELAND APPROXIMATION

When the geometrical extent of the plasma, l_{pl}, is large or the density is high, the mean free path of photons is small in comparison with l_{pl}

$$l_\nu \cong 1 / \varkappa'_\nu \gg l_{pl}$$

In this case, the radiation "diffuses" to the outer plasma regions (diffusion approximation). The radiation transport equation

$$\nabla S_\nu = c\, \varkappa'_\nu \left(u_\nu - \rho_\nu\right)$$

leads to the monochromatic flux vector

$$S_\nu \cong - \frac{c}{3\, \varkappa'_\nu} \nabla u_\nu = - \varkappa_\nu^{rad}\, \nabla T$$

with the spectral coefficient of radiation heat transport

$$\varkappa_\nu^{rad} = \frac{c}{3\, \varkappa'_\nu} \frac{\partial u_\nu}{\partial T}$$

Integration over frequency gives

$$S = -\frac{c}{3\,\bar{\varkappa}_{Ross}} \int_0^\infty \frac{\partial u_\nu}{\partial t}\, d\nu \;\; \nabla T = -\varkappa^{rad}\, \nabla T$$

with the Rosseland mean absorption coefficient

$$\bar{\varkappa}_{Ross} = \frac{\int_0^\infty \partial u_\nu/\partial T \, d\nu}{\int_0^\infty \frac{1}{\varkappa'_\nu}\, \partial u_\nu/\partial T \, d\nu}$$

and the (mean) coefficient of radiation heat transport

$$\varkappa^{rad} = \frac{16}{3} \frac{\sigma_{St/B}}{\bar{\varkappa}_{Ross}} T^3$$

where $\sigma_{St/B}$ is the → Stefan-Boltzmann constant.

The units of these quantities are

	u_ν and ρ_ν	u	S_ν	S	$\bar{\varkappa}_{Planck}$ and $\bar{\varkappa}_{Ross}$
SI	J/m^3 Hz	J/m^3	W/m^2 Hz	W/m^2	m^{-1}
CGS	erg/cm^3 Hz	erg/cm^3	$erg/s\ cm^2$ Hz	$erg/s\ cm^2$	cm^{-1}

	$\varkappa$ and $\varkappa'$	$\varkappa_\nu^{rad}$	$\varkappa^{rad}$	l_ν	l_{pl}	τ
SI	m^{-1}	W/K m Hz	W/K m	m	m	-
CGS	cm^{-1}	erg/s K cm Hz	erg/s K cm	cm	cm	-

References: [48], [58], [191]-[194]

11. HIGH-TEMPERATURE PLASMA PHYSICS AND THERMONUCLEAR FUSION

11.1. THERMONUCLEAR REACTIONS

Fusion reactions are:

$D + D \longrightarrow T\ (1.01\ MeV) + p\ (3.02\ MeV)$	50%	(1a)
$ \longrightarrow He^3\ (0.82\ MeV) + n\ (2.45\ MeV)$	50%	(1b)
$D + T \longrightarrow He^4\ (3.5\ MeV) + n\ (14.1\ MeV)$		(2)
$D + He^3 \longrightarrow He^4\ (3.6\ MeV) + p\ (14.7\ MeV)$		(3)
$T + T \longrightarrow He^4 + 2\ n + 11.3\ MeV$		(4)
$He^3 + T \longrightarrow He^4 + n + p + 12.1\ MeV$	51%	(5a)
$ \longrightarrow He^4\ (4.8\ MeV) + D\ (9.5\ MeV)$	43%	(5b)
$ \longrightarrow He^5\ (2.4\ MeV) + p\ (11.9\ MeV)$	6%	(5c)
$p + Li^6 \longrightarrow He^4\ (1.7\ MeV) + He^3\ (2.3\ MeV)$		(6)
$p + Li^7 \longrightarrow 2\ He^4 + 17.3\ MeV$	nearly 20%	(7a)
$ \longrightarrow Be^7 + n - 1.6\ MeV$	nearly 80%	(7b)
$D + Li^6 \longrightarrow 2\ He^4 + 22.4\ MeV$		(8)
$p + B^{11} \longrightarrow 3\ He^4 + 8.7\ MeV$		(9)
$n + Li^6 \longrightarrow He^4\ (2.1\ MeV) + T\ (2.7\ MeV)$		(10)

The total-cross section as a function of E, the energy of the incident particle (the first ion on the left side of (1)-(5)) assuming the target ion at rest, can be fitted by the expression

$$\sigma_T(E) = \frac{A_5 + \left[(A_4 - A_3 E)^2 + 1\right]^{-1} A_2}{E\left[\exp\left(A_1 E^{-1/2}\right) - 1\right]}$$

Here σ is in barns (1 barn = 10^{-24} cm^2 = 10^{-28} m^2) and E is in keV.

The Duane coefficients, A_j, for the main fusion reactions are

	D - D (1a)	D - D (1b)	D - T (2)	D - He^3 (3)	T - T (4)	T - He^3 (5a-c)
A_1	46.097	47.88	45.95	89.27	38.39	123.1
A_2	372	482	50200	25900	448	11250
A_3	0.000436	0.000308	0.01368	0.00398	0.00102	0
A_4	1.220	1.177	1.076	1.297	2.09	0
A_5	0	0	409	647	0	0

The corresponding reaction rates (in cm^3/s) averaged over → Maxwellian distributions, i.e.

$$\langle\sigma v\rangle = \int \sigma\ v\ f_M\ dv$$

are

temperature in keV	D - D (1a + 1b)	D - T (2)	D - He^3 (3)	T - T (4)	T - He^3 (5a-c)
1.0	1.5×10^{-22}	5.5×10^{-21}	10^{-26}	3.3×10^{-22}	10^{-28}
2.0	5.4×10^{-21}	2.6×10^{-19}	1.4×10^{-23}	7.1×10^{-21}	10^{-25}
5.0	1.8×10^{-19}	1.3×10^{-17}	6.7×10^{-21}	1.4×10^{-19}	2.1×10^{-22}
10.0	1.2×10^{-18}	1.1×10^{-16}	2.3×10^{-19}	7.2×10^{-19}	1.2×10^{-20}
20.0	5.2×10^{-18}	4.2×10^{-16}	3.8×10^{-18}	2.5×10^{-18}	2.6×10^{-19}
50.0	2.1×10^{-17}	8.7×10^{-16}	5.4×10^{-17}	8.7×10^{-18}	5.3×10^{-18}
100.0	4.5×10^{-17}	8.5×10^{-16}	1.6×10^{-16}	1.9×10^{-17}	2.7×10^{-17}
200.0	8.8×10^{-17}	6.3×10^{-16}	2.4×10^{-16}	4.2×10^{-17}	9.2×10^{-17}
500.0	1.8×10^{-16}	3.7×10^{-16}	2.3×10^{-16}	8.4×10^{-17}	2.9×10^{-16}
1000.0	2.2×10^{-16}	2.7×10^{-16}	1.8×10^{-16}	8.0×10^{-17}	5.2×10^{-16}

Branching of DD reactions:

branches	D - temperature in keV 10	20	50	100	reaction energy
DD n ^3He	51.0 %	52.0 %	53.7 %	54.9 %	3.27 MeV
DD p T	49.0 %	48.0 %	46.3 %	45.1 %	4.03 MeV

For low energies, $T \leq 25$ keV, the data can be represented by

$$\langle\sigma v\rangle_{DD} = 2.33 \times 10^{-14}\ T^{-2/3}\ \exp\left(-\ 18.78\ T^{-1/3}\right)$$

$$\langle\sigma v\rangle_{DT} = 3.68 \times 10^{-12}\ T^{-2/3}\ \exp\left(-\ 19.94\ T^{-1/3}\right)$$

where $\langle\sigma v\rangle$ is in cm^3/s and T in keV.

The power density released in the form of charged particles is

$$P_{DD} = \begin{cases} 3.3 \times 10^{-19} \left(n_D\right)^2 \langle\sigma v\rangle_{DD} & (SI) \\ 3.3 \times 10^{-13} \left(n_D\right)^2 \langle\sigma v\rangle_{DD} & (CGS) \end{cases}$$

(including the subsequent D-T reaction)

$$P_{DT} = \begin{cases} 5.6 \times 10^{-19}\ n_D\ n_T\ \langle\sigma v\rangle_{DT} & (SI) \\ 5.6 \times 10^{-13}\ n_D\ n_T\ \langle\sigma v\rangle_{DT} & (CGS) \end{cases}$$

$$P_{DHe^3} = \begin{cases} 2.9 \times 10^{-18}\ n_D\ n_{He^3}\ \langle\sigma v\rangle_{DT} & (SI) \\ 2.9 \times 10^{-12}\ n_D\ n_{He^3}\ \langle\sigma v\rangle_{DT} & (CGS) \end{cases}$$

<u>References</u>: [35], [157]

11.2. FUSION PRODUCT AND RESULTS ACHIEVED BY THE LARGEST TOKAMAKS

The following figure illustrates, in ($n\tau$), T_i space the performance, since 1960, of the world's largest tokamaks. T_i is the central ion temperature and $n\tau$ the confinement parameter. Data points are extrapolated values based on the performance which would have been achieved if equal densities of D and T had been used as fuel ($T_i = T_e$ assumed, temperature profiles assumed flat).

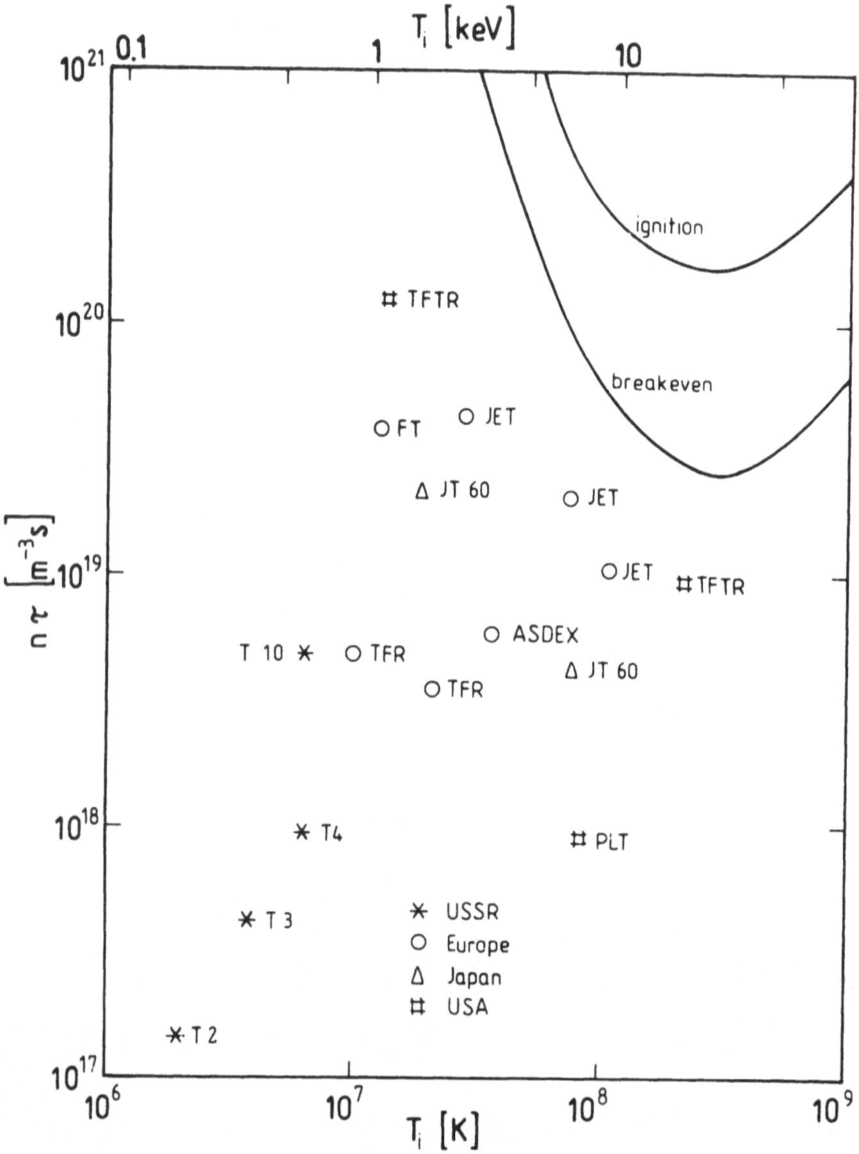

Several criteria corresponding to different values of $n\tau$ can be distinguished:

(1) breakeven is reached when the fusion power is equal to the input power. Thermonuclear conditions are then maintained without external power input so long as fuel is supplied.

(2) ignition is reached if the power absorbed by particles produced during the reaction (1/5 of the fusion power) is equal to the power lost due to radiation, conduction and convection.

(3) Lawson criterion is fulfilled when recovery of the energy produced by fusion reactions (with an assumed efficiency in converting heat to electrical energy of about 30%) provides all the energy required to sustain thermonuclear conditions. The Lawson criterion lies between breakeven and ignition.

References: [60], [128]-[131]

12. FEATURES OF ATOMS AND IONS

12.1. SPECTROSCOPIC NOTATION

Quantum numbers (angular momenta are measured in units of $\hbar = h/2\pi$)

j	electron angular momentum; $j = l + s$
J	atomic angular momentum
l	electron orbital angular momentum; $l = 0, 1, 2, \ldots$
L	atomic orbital angular momentum
m_l	electron magnetic orbital angular momentum (z-component of l); $m_l = l,\ l-1,\ l-2, \ldots, -l$
M	atomic magnetic orbital angular momentum
s	electronic spin; $s = \pm 1/2$
S	atomic spin
m_s	electron magnetic spin angular momentum (z-component of s); $m_s = \pm 1/2$
n	principal quantum number

For a prescribed value l, it is customary to number the states of a particle in ascending order of energy by the principal quantum number n. In complex atoms, the sequence of energy levels differ from that of hydrogen. In hydrogen, E depends on n and not on l, whilst $E_{n+1} > E_n$ is always true.
The distribution of electrons in an atom with respect to states with different values of n and l is called the electronic configuration. If there are several electrons (k in number) with the same values of n and l, this is denoted by

$$\boxed{n\, l^k}$$

It is usual to denote states with different values of l by letters of the alphabet as follows

$l =$	0,	1,	2,	3,	4,	5,	6
	s	p	d	f	g	h	i

A complete characterization of states of an electron is achieved by the assignment of n, l, m_l, m_s, the energy level being defined only by n and l in so far as the central field approximation is valid. For a given l, m_l can take $2l+1$ values, while m_s can be $\pm 1/2$. There are a total of $2(2l+1)$ states with the same value of n and l (equivalent states).

Pauli's principle: There can be no more than one electron in each n, l, m_l, m_s state.

An assembly of $2(2l+1)$ equivalent electrons is a sub-shell or orbital:

l	$2(2l+1)$	name
0	2	s orbital
1	6	p orbital
2	10	d orbital
3	14	f orbital
4	18	g orbital
5	22	h orbital

An electron shell is defined by

n	name	states	maximum number of electrons, $2n^2$
1	K shell	1s	2
2	L shell	2s, 2p	8
3	M shell	3s, 3p, 3d	18
4	N shell	4s, 4p, 4d, 4f	32
5	O shell	5s, 5p, 5d, 5f, 5g	50
6	P shell	6s, 6p, 6d, 6f, 6g, 6h	72

Electrostatic interaction between electrons and spin-orbit interactions lead to splitting of the levels n_1l_1, n_2l_2,... into a number of sub-levels (a violation of the approximation of central symmetry), characterized by different values of the total orbital momentum, L, and the total spin, S.
Russel-Saunders coupling (R-S coupling, LS coupling): the electrostatic interaction is much stronger than the spin-orbit interaction.
Otherwise: jj-coupling.

LS COUPLING APPROXIMATION

Hund's rule: The level with the greatest possible value of S for a given electronic configuration and the greatest possible (for this S) value of L has the lowest energy.
Energy levels corresponding to a definite value of L and S are called spectral

terms or simply terms. It is customary to denote the values of L as follows

L =	0,	1,	2,	3,	4,	5,	6,	7,	8,	9,	10
	S	P	D	F	G	H	I	K	L	M	N

The total angular momentum of an atom can take the values

$$L + S \geq J \geq |L - S|$$

If $L \geq S$, then $2S+1$ different values of J are possible. The number $2S+1$ determines the number of components of the term (multiplicity, $2S+1=1$: singlet, 2: doublet, 3: triplet, 4:quartet, etc.;fine structure of terms).
The full designation of a term has the form

$$\boxed{{}^{2S+1}L_J^{\circ}}$$

Here, the parity $(-1)^{l_1} (-1)^{l_2} (-1)^{l_3} \ldots$ of a given term is denoted by the superscript "∘" (odd), placed to the top right of L. Absence of "∘" indicates that the term has even parity.
In the terms LS there are

$$(2L+1)\ (2S+1)$$

states, differing by values of the z-components M_L and M_S of the total orbital and spin angular momenta L and S, respectively.
The statistical weight of the level J is $(2J+1)$. The statistical weight of the term LS is $(2L+1)(2S+1)$. The statistical weight of the configuration is given by $\sum_J (2J+1) = (2L+1)\ (2S+1)$.

Terms of the l^k configuration:

configuration	terms		statistical weight of configuration
s	^{2}S		2
s^2	^{1}S		1
p or p^5	^{2}P$^\circ$		6
p^2 or p^4	^{1}S, ^{1}D	^{3}P	15
p^3	^{2}P, ^{2}D	^{4}S$^\circ$	20
d or d^9	^{2}D		10
d^2 or d^8	^{1}S, ^{1}D, ^{1}G	^{3}P, ^{3}F	45
⋮			

jj COUPLING APPROXIMATION

In this case, spin-orbit interaction is large and the concept of separate orbital and spin angular momenta of an electron loses meaning. *jj* coupling is of interest for heavy elements and multiply charged ions.

In the *jj* coupling scheme, the state of each electron is determined by $n\ l\ j\ m_j$. For a given of j, $l = j \pm 1/2$. It is conventional to write this as

$$\boxed{l_j}$$

The following states are possible:

$s_{1/2}$, $p_{1/2}$, $p_{3/2}$, $d_{3/2}$, $d_{5/2}$, $f_{5/2}$, $f_{7/2}$, $g_{7/2}$, $g_{9/2}$, $h_{9/2}$, $h_{11/2}$,...

the parity of the states s, d, g,... being even, and of the states p, f, h, ... odd.

If electrostatic interaction is completely neglected, the energy of each electron does not depend on the orientation of j. Their state is completely specified by $n\ l\ j$ with a $(2j+1)$-fold degeneration.

If electrostatic interaction is taken into account, a level described by $n_i l_i j_i$, related to each electron, splits into sub-levels which are characterized by a value of J. Example: a given configuration has *np nd*. For a "p" electron, j= 1/2, 3/2; for a "d" electron, j= 3/2, 5/2. States with given j_1, j_2, and J are denoted by

$$\boxed{(j_1\ j_2)_J}$$

where $|j_1 - j_2| \le J \le j_1 + j_2$. In this example, the states with j_1=1/2, j_2=3/2 are $(1/2\ \ 3/2)_1$ and $(1/2\ \ 3/2)_2$.

jl COUPLING OF HIGHLY EXCITED ATOMS

If the spin-orbit interaction of electrons in an atomic core exceeds the electrostatic interaction between the valence electron and the electrons of the atomic core, the core can be described using the values of L and S whilst the interaction with the valence electron can be treated using the *jj* coupling approximation. The state *LSjl* splits into a set of levels which differ in quantum number

$$K = j + l$$

when the electrostatic interaction of the valence electron with the core is taken into account. If, additionally, spin-orbit interaction is taken into account, each level *LSjlK* splits into two sub-levels characterized by

$$J = K \pm 1/2$$

The system can then be summarized with the following notation

$$\boxed{{}^{2S+1}L_j \; n \; l \; [K]_J}$$

where *n* is the principal quantum number of the valence electron.

12.2. ELECTRONIC STRUCTURE OF ATOMS AND IONS

In the table below, the following are compiled

* Z : nuclear charge of the element (EL.)
* m : atomic mass, in atomic mass units, averaged according to isotopic distribution
* configuration of valence electrons and term of the ground state of the neutral atom (I), of the singly ionized ion (II), of the doubly ionized ion (III), and of the three-fold ionized ion (IV)
* ionization energy, in eV, of the neutral atom, and of the singly, doubly, and three-fold ionized ion. These values are given below the respective spectroscopic terms

Z	El.	m in a.m.u.	I	II	III	IV
1	H	1.007825	1s $-{}^2S_{1/2}$ 13.598			
	D	2.014102	1s $-{}^2S_{1/2}$ 13.602			
	T	3.016049	1s $-{}^2S_{1/2}$ 13.603			
2	He	4.0026	1s $-{}^1S_0$ 24.587	1s $-{}^2S_{1/2}$ 54.418		
3	Li	6.94	2s $-{}^2S_{1/2}$ 5.392	1s $-{}^1S_0$ 75.641	1s $-{}^2S_{1/2}$ 122.45	
4	Be	9.01218	$2s^2$ $-{}^1S_0$ 9.323	2s $-{}^2S_{1/2}$ 18.211	$1s^2$ $-{}^1S_0$ 153.90	1s $-{}^2S_{1/2}$ 217.72
5	B	10.81	2p $-{}^2P_{1/2}$ 8.298	$2s^2$ $-{}^1S_0$ 25.155	2s $-{}^2S_{1/2}$ 37.93	$1s^2$ $-{}^1S_0$ 259.38
6	C	12.011	$2p^2$ $-{}^3P_0$ 11.260	2p $-{}^2P^o_{1/2}$ 24.384	$2s^2$ $-{}^1S_0$ 47.89	2s $-{}^2S_{1/2}$ 64.49

Z	El.	m in a.m.u.	I	II	III	IV
7	N	14.0067	$2p^3\ -{}^4P^o_{3/2}$ 14.534	$2p^2\ -{}^3P_0$ 29.602	$2p\ -{}^2P^o_{1/2}$ 47.45	$2s^2\ -{}^1S_0$ 77.47
8	O	15.9994	$2p^4\ -{}^3P_2$ 13.618	$2p^3\ -{}^4S^o_{3/2}$ 35.118	$2p^2\ -{}^3P_0$ 54.94	$2p\ -{}^2P^o_{1/2}$ 77.41
9	F	18.9984	$2p^5\ -{}^2P^0_{3/2}$ 17.423	$2p^4\ -{}^3P_2$ 34.971	$2p^3\ -{}^4S^o_{3/2}$ 62.71	$2p^2\ -{}^3P_0$ 87.14
10	Ne	20.179	$2p^6\ -{}^1S_0$ 21.565	$2p^5\ -{}^2P^o_{3/2}$ 40.963	$2p^4\ -{}^3P_2$ 63.46	$2p^3\ -{}^4S^o_{3/2}$ 97.12
11	Na	22.9898	$3s\ -{}^2S_{1/2}$ 5.139	$2p^6\ -{}^1S_0$ 47.287	$2p^5\ -{}^2P^o_{3/2}$ 71.62	$2p^4\ -{}^3P_2$ 98.92
12	Mg	24.305	$3s^2\ -{}^1S_0$ 7.646	$3s\ -{}^2S_{1/2}$ 15.035	$2p^6\ -{}^1S_0$ 80.14	$2p^5\ -{}^2P^o_{3/2}$ 109.27
13	Al	26.982	$3p\ -{}^2P^o_{1/2}$ 5.986	$3s^2\ -{}^1S_0$ 18.829	$3s\ -{}^2S_{1/2}$ 28.45	$2p^6\ -{}^1S_0$ 119.99
14	Si	28.086	$2p^2\ -{}^3P_0$ 8.151	$3p\ -{}^2P^o_{1/2}$ 16.346	$3s^2\ -{}^1S_0$ 33.49	$3s\ -{}^2S_{1/2}$ 45.14
15	P	30.9738	$3p^3\ -{}^4S^o_0$ 10.487	$3p^2\ -{}^3P_0$ 19.770	$3p\ -{}^2P^o_{1/2}$ 30.20	$3s^2\ -{}^1S_0$ 51.44
16	S	32.07	$3p^4\ -{}^3P_2$ 10.360	$3p^3\ -{}^4S^o_{3/2}$ 23.338	$3p^2\ -{}^3P_0$ 34.83	$3p\ -{}^2P^o_{1/2}$ 47.31
17	Cl	35.453	$3p^5\ -{}^2P^o_{3/2}$ 12.968	$3p^4\ -{}^3P_2$ 23.814	$3p^3\ -{}^4S^o_{3/2}$ 39.61	$3p^2\ -{}^3P_0$ 53.47
18	Ar	39.948	$3p^6\ -{}^1S_0$ 15.760	$3p^5\ -{}^2P^o_{3/2}$ 27.630	$3p^4\ -{}^3P_2$ 40.91	$3p^3\ -{}^4S_{3/2}$ 59.81
19	K	39.098	$4s\ -{}^2S_{1/2}$ 4.341	$3p^6\ -{}^1S_0$ 31.63	$3p^5\ -{}^2P^o_{3/2}$ 45.81	$3p^4\ -{}^3P_2$ 60.91
20	Ca	40.078	$4s^2\ -{}^1S_0$ 6.113	$4s\ -{}^2S_{1/2}$ 11.872	$3p^6\ -{}^1S_0$ 50.91	$3p^5\ -{}^2P^o_{3/2}$ 67.3
21	Sc	44.9559	$3d\ 4s^2-{}^2D_{3/2}$ 6.562	$3d\ 4s\ -{}^3D_1$ 12.800	$3d\ -{}^2D_{3/2}$ 24.76	$3p^6\ -{}^1S_0$ 73.49
22	Ti	47.88	$3d^24s^2-{}^3F_2$ 6.82	$3d^24s\ -{}^4F_{3/2}$ 13.58	$3d^2\ -{}^3F_2$ 27.49	$3d\ -{}^2D_{3/2}$ 43.27
23	V	50.9415	$3d^34s^2-{}^4F_{3/2}$ 6.74	$3d^4\ -{}^5D_0$ 14.66	$3d^3\ -{}^4F_{3/2}$ 29.31	$3d^2\ -{}^3F_2$ 46.71
24	Cr	51.996	$3d^54s\ -{}^7S_3$ 6.766	$3d^5\ -{}^6S_{5/2}$ 16.50	$3d^4\ -{}^5D_0$ 31.0	$3d^3\ -{}^4F_{3/2}$ 49.2

Z	El.	m in a.m.u.	I	II	III	IV
25	Mn	54.938	$3d^5 4s^2 - {}^6S_{5/2}$ 7.434	$3d^5 4s - {}^7S_3$ 15.640	$3d^5 - {}^6S_{5/2}$ 33.67	$3d^4 - {}^5D_0$ 51.2
26	Fe	55.847	$3d^6 4s^2 - {}^5D_4$ 7.902	$3d^6 4s - {}^6D_{9/2}$ 16.188	$3d^6 - {}^5D_4$ 30.65	$3d^5 - {}^6S_{5/2}$ 54.8
27	Co	58.933	$3d^7 4s^2 - {}^4F_{9/2}$ 7.86	$3d^8 - {}^3F_4$ 17.084	$3d^7 - {}^4F_{9/2}$ 33.5	$3d^6 - {}^5D_4$ 51.3
28	Ni	58.69	$3d^8 4s^2 - {}^3F_4$ 7.637	$3d^9 - {}^2D_{5/2}$ 18.169	$3d^8 - {}^3F_4$ 35.3	$3d^7 - {}^4F_{9/2}$ 54.9
29	Cu	63.55	$4s - {}^2S_{1/2}$ 7.726	$3d^{10} - {}^1S_0$ 20.293	$3d^9 - {}^2D_{5/2}$ 36.84	$3d^8 - {}^3F_4$ 57.4
30	Zn	65.39	$4s^2 - {}^1S_0$ 9.394	$4s - {}^2S_{1/2}$ 17.964	$3d^{10} - {}^1S_0$ 39.72	$3d^9 - {}^2D_{5/2}$ 59.57
31	Ga	69.72	$4p - {}^2P^o_{1/2}$ 5.999	$4s^2 - {}^1S_0$ 20.515	$4s - {}^2S_{1/2}$ 30.7	$3d^{10} - {}^1S_0$ 64.2
32	Ge	72.59	$4p^2 - {}^3P_0$ 7.899	$4p - {}^2P^o_{1/2}$ 15.935	$4s^2 - {}^1S_0$ 34.2	$4s - {}^2S_{1/2}$ 45.7
33	As	74.9216	$4p^3 - {}^4S^o_{3/2}$ 9.789	$4p^2 - {}^3P_0$ 18.59	$4p - {}^2P^o_{1/2}$ 28.4	$4s^2 - {}^1S_0$ 50.1
34	Se	78.96	$4p^4 - {}^3P_2$ 9.752	$4p^3 - {}^4S^o_{3/2}$ 21.16	$4p^2 - {}^3P_0$ 30.82	$4p - {}^2P^o_{1/2}$ 42.95
35	Br	79.904	$4p^5 - {}^2P^o_{3/2}$ 11.814	$4p^4 - {}^3P_2$ 21.81	$4p^3 - {}^4S^o_{3/2}$ 35.90	$4p^2 - {}^3P_0$ 47.3
36	Kr	83.8	$4p^6 - {}^1S_0$ 14.000	$4p^5 - {}^2P^o_{3/2}$ 24.360	$4p^4 - {}^3P_2$ 36.95	$4p^3 - {}^4S^o_{3/2}$ 52.5
37	Rb	85.468	$5s - {}^2S_{1/2}$ 4.177	$4p^6 - {}^1S_0$ 27.290	$4p^5 - {}^2P^o_{3/2}$ 39.2	$4p^4 - {}^3P_2$ 52.6
38	Sr	87.62	$5s^2 - {}^1S_0$ 5.695	$5s - {}^2S_{1/2}$ 11.030	$4p^6 - {}^1S_0$ 42.88	$4p^5 - {}^2P^o_{5/2}$ 56.28
39	Y	88.9059	$4d\,5s^2 - {}^2D_{3/2}$ 6.217	$5s^2 - {}^1S_0$ 12.24	$5s - {}^2S_{1/2}$ 20.53	$4p^6 - {}^1S_0$ 60.61
40	Zr	91.224	$4d^2 5s^2 - {}^3F_2$ 6.837	$4d^2 5s - {}^4F_{3/2}$ 13.13	$4d^2 - {}^3F_2$ 23.1	$4d - {}^2D_{3/2}$ 34.42
41	Nb	92.90638	$4d^4 5s - {}^6D^o_{1/2}$ 6.88	$4d^4 - {}^5D_0$ 14.32	$4d^3 - {}^4F_{3/2}$ 25.0	$4d^2 - {}^3F_2$ 37.7
42	Mo	95.94	$4d^5 5s - {}^7S_3$ 7.099	$4d^5 - {}^6S_{5/2}$ 16.16	$4d^4 - {}^5D_0$ 27.2	$4d^3 - {}^4F_{3/2}$ 46.4

Z	El.	m in a.m.u.	I	II	III	IV
43	Tc	(98)	$4d^5 5s^2 - {}^6S_{5/2}$ 7.28	$4d^5 5s - {}^7S_3$ 15.26	$4d^5 - {}^6S_{5/2}$ 29.5	$4d^4 - {}^5D_0$ -
44	Ru	101.1	$4d^7 5s - {}^5F_5$ 7.366	$4d^7 - {}^4F_{9/2}$ 16.76	$4d^6 - {}^5D_4$ 28.5	$4d^5 - {}^6S_{5/2}$ -
45	Rh	102.9055	$4d^8 5s - {}^3F_{3/2}$ 7.46	$4d^8 - {}^3F_4$ 18.08	$4d^7 - {}^4F_{9/2}$ 31.1	$4d^6 - {}^5D_4$ -
46	Pd	106.4	$4d^{10} - {}^1S_0$ 8.336	$4d^9 - {}^2D_{5/2}$ 19.43	$4d^8 - {}^3F_4$ 32.9	$4d^7 - {}^4F_{9/2}$ -
47	Ag	107.868	$5s - {}^2S_{1/2}$ 7.576	$4d^{10} - {}^1S_0$ 21.49	$4d^9 - {}^2D_{5/2}$ 34.8	$4d^8 - {}^3F_4$ -
48	Cd	112.4	$5s^2 - {}^1S_0$ 8.994	$5s - {}^2S_{1/2}$ 16.908	$4d^{10} - {}^1S_0$ 37.47	$4d^9 - {}^2D_{5/2}$ -
49	In	114.8	$5s^2 5p - {}^2P^o_{1/2}$ 5.786	$5s^2 - {}^1S_0$ 18.87	$5s - {}^2S_{1/2}$ 28.0	$4d^{10} - {}^1S_0$ 57.0
50	Sn	118.71	$5p^2 - {}^3P_0$ 7.344	$5s^2 5p - {}^2P_{1/2}$ 14.632	$5s^2 - {}^1S_0$ 30.50	$5s - {}^2S_{1/2}$ 40.74
51	Sb	121.8	$5p^3 - {}^4S^o_{3/2}$ 8.609	$5s^2 5p^2 - {}^3P_0$ 16.53	$5s^2 5p - {}^2P^o_{1/2}$ 25.32	$5s^2 - {}^1S_0$ 44.16
52	Te	127.60	$5p^4 - {}^3P_2$ 9.010	$5p^3 - {}^4S^o_{3/2}$ 18.6	$5p^2 - {}^3P^o_0$ 27.96	$5s^2 5p - {}^2P^o_{1/2}$ 37.42
53	I	126.9045	$5p^5 - {}^2P^o_{3/2}$ 10.451	$5p^4 - {}^3P_2$ 19.131	$5p^3 - {}^4S^o_{3/2}$ 33.0	$5p^2 - {}^3P_0$ -
54	Xe	131.29	$5p^6 - {}^1S_0$ 12.130	$5p^5 - {}^2P^o_{3/2}$ 20.98	$5p^4 - {}^3P_2$ 31.0	$5p^3 - {}^4S^o_{3/2}$ 45
55	Cs	132.9054	$6s - {}^2S_{1/2}$ 3.894	$5p^6 - {}^1S_0$ 23.15	$5p^5 - {}^2P^o_{3/2}$ 33.4	$5p^4 - {}^3P_2$ 46
56	Ba	137.33	$6s^2 - {}^1S_0$ 5.212	$6s - {}^2S_{1/2}$ 10.00	$5p^6 - {}^1S_0$ 35.8	$5p^5 - {}^2P^o_{3/2}$ 47
57	La	138.9055	$5d\ 6s^2 - {}^2D_{3/2}$ 5.577	$5d^2 - {}^3F_2$ 11.1	$5d - {}^2D_{3/2}$ 19.18	$5p^6 - {}^1S_0$ 49.9
58	Ce	140.12	$4f5d6s^2 - {}^1G^o_4$ 5.539	$4f5d^2 - {}^4H^o_{7/2}$ 10.8	$4f^2 - {}^3H_4$ 20.20	$4f - {}^2F_{5/2}$ 36.76
59	Pr	140.9077	$4f^3 6s^2 - {}^4I^o_{9/2}$ 5.47	$4f^3\ ({}^4I^o_{9/2})$ $6s\ \left(\frac{9}{2}, \frac{1}{2}\right)^o_4$ 10.6	$4f^3 - {}^4I^o_{9/2}$ 21.62	$4f^2 - {}^3H_4$ 38.98

Z	El.	m in a.m.u.	I	II	III	IV
60	Nd	144.24	$4f^4 6s^2 - {}^5I_4$	$4f^4 6s - {}^6I_{7/2}$	$4f^4 - {}^5I_4$	$4f^3 - {}^4I^o_{9/2}$
			5.525	10.7	22.1	40.4
61	Pm	(145)	$4f^5 6s^2 - {}^6H^o_{5/2}$	$4f^5 6s - {}^7H^o_2$	$4f^5 - {}^6H^o_{5/2}$	$4f^4 - {}^5I_4$
			5.58	10.9	22.3	41.0
62	Sm	150.36	$4f^6 6s^2 - {}^7F_0$	$4f^6 6s - {}^8F_{1/2}$	$4f^6 - {}^7F_0$	$4f^5 - {}^6H^o_{5/2}$
			5.644	11.1	23.4	41.4
63	Eu	151.96	$4f^7 6s^2 - {}^8S^o_{7/2}$	$4f^7 6s - {}^9S^o_4$	$4f^7 - {}^8S^o_{7/2}$	$4f^6 - {}^7F_0$
			5.670	11.24	24.9	42.7
64	Gd	157.25	$4f^7 5d6s^2 - {}^9D^o_2$	$4f^7 5d6s - {}^{10}D^o_{\frac{5}{2}}$	$4f^7 5d - {}^9D^o_2$	$4f^7 - {}^8S^o_{7/2}$
			6.150	12.1	20.6	44.0
65	Tb	158.9254	$4f^9 6s^2 - {}^6H^o_{\frac{15}{2}}$	$4f^9 ({}^6H^o_{15/2})$ $6s \left(\frac{15}{2}, \frac{1}{2}\right)^o_8$	$4f^9 - {}^6H^o_{15/2}$	$4f^8 - {}^7F_6$
			5.864	11.5	21.9	39.4
66	Dy	162.5	$4f^{10} 6s^2 - {}^5I_8$	$4f^{10} ({}^5I_8)$ $6s \ (8, \frac{1}{2})_{17/2}$	$4f^{10} - {}^5I_8$	$4f^9 - {}^6H^o_{15/2}$
			5.939	11.7	22.8	41.4
67	Ho	164.9304	$4f^{11} 6s^2 - {}^4I^o_{\frac{15}{2}}$	$4f^{11} \left({}^4I^o_{15/2}\right)$ $6s \left(\frac{15}{2}, \frac{1}{2}\right)^o_8$	$4f^{11} - {}^4I^o_{15/2}$	$4f^{10} - {}^5I^o_8$
			6.022	11.8	22.8	42.5
68	Er	167.26	$4f^{12} 6s^2 - {}^3H_6$	$4f^{12} ({}^3H_6)$ $6s \ (8, \frac{1}{2})_{17/2}$	$4f^{12} - {}^3H_6$	$4f^{11} - {}^4I^o_{15/2}$
			6.108	11.9	22.7	42.7
69	Tm	168.9342	$4f^{13} 6s^2 - {}^2F^o_{\frac{7}{2}}$	$4f^{13} ({}^2F^o_{7/2})$ $6s \left(\frac{7}{2}, \frac{1}{2}\right)^o_4$	$4f^{13} - {}^2F^o_{7/2}$	$4f^{12} - {}^3H_6$
			6.184	12.1	23.7	42.7
70	Yb	173.04	$4f^{14} 6s^2 - {}^1S_0$	$4f^{14} 6s - {}^2S_{1/2}$	$4f^{14} - {}^1S_0$	$4f^{13} - {}^2F^o_{7/2}$
			6.254	12.18	25.05	43.6
71	Lu	174.967	$4f^{14} 5d6s^2 - {}^2D_{\frac{3}{2}}$	$4f^{14} 6s^2 - {}^1S_0$	$4f^{14} 6s - {}^2S_{\frac{1}{2}}$	$4f^{14} - {}^1S_0$
			5.426	13.9	20.96	45.25
72	Hf	178.49	$5d^2 6s^2 - {}^3F_2$	$5d \ 6s^2 - {}^2D_{3/2}$	$5d^2 - {}^3F_2$	$5d - {}^2D_{3/2}$
			6.8	14.9	23.3	33.4

Z	El.	m in a.m.u.	I	II	III	IV
73	Ta	180.948	$5d^3 6s^2 - {}^4F_{3/2}$ 7.89	$5d^3 6s - {}^5F_1$		
74	W	183.85	$5d^4 6s^2 - {}^5D_0$ 7.98	$5d^4 6s - {}^6D_{1/2}$		
75	Re	186.207	$5d^5 6s^2 - {}^6S_{5/2}$ 7.88	$5d^5 6s - {}^7S_3$		
76	Os	190.2	$5d^6 6s^2 - {}^5D_4$ 8.73	$5d^6 6s - {}^6D_{9/2}$		
77	Ir	192.22	$5d^7 6s^2 - {}^4F_{9/2}$ 9.05	$5d^7 6s - {}^5F_5$		
78	Pt	195.08	$5d^9 6s - {}^3D_3$ 8.96	$5d^9 - {}^2D_{5/2}$ 18.56		
79	Au	196.9665	$5d^{10} 6s - {}^2S_{1/2}$ 9.226	$5d^{10} - {}^1S_0$ 20.5	$5d^9 - {}^2D_{5/2}$ 34	$5d^8 - {}^3F$ 43
80	Hg	200.59	$6s^2 - {}^1S_0$ 10.438	$5d^{10} 6s - {}^2S_{1/2}$ 18.76	$5d^{10} - {}^1S_0$ 34.2	$5d^9 - {}^2D_{5/2}$ 46
81	Tl	204.383	$6s^2 6p - {}^2P^o_{1/2}$ 6.101	$6s^2 - {}^1S_0$ 20.43	$6s - {}^2S_{1/2}$ 29.85	
82	Pb	207.2	$6s^2 6p^2 - {}^3P_0$ 7.417	$6s^2 6p - {}^2P^o_{1/2}$ 15.03	$6s^2 - {}^1S^o_0$ 31.94	$6s - {}^2S_{1/2}$ 42.33
83	Bi	208.9804	$6s^2 6p^3 - {}^4S^o_{3/2}$ 7.286	$6p^2 - {}^3P_0$ 16.7	$6s^2 6p - {}^2P^o_{\frac{1}{2}}$ 25.56	$6s^2 - {}^1S_0$ 45.3
90	Th	232.038	$6d^2 7s^2 - {}^3F_2$ 6.1	$6d^2 7s - {}^4F_{3/2}$ 11.9	$5f\ 6d - {}^3H^o_4$ 18.3	$5f - {}^2F^o_{5/2}$ 28.7
92	U	238.03	$5f^3\left({}^4I^o_{9/2}\right)$ $6d7s\ \left(\frac{9}{2},\frac{3}{2}\right)^o_6$ 6.194	$5f^3 7s^2 - {}^4I^o_{9/2}$ 11.9	20	$5f^3 - {}^4I^o_{9/2}$ 37

References: [118]-[121]

12.3. RADIATIVE TRANSITIONS

Radiative transitions are governed by the following <u>selection rules</u>:

* Multipole radiation of the order $\varkappa$ satifies the selection rules:

$$|\Delta J| = |J' - J| = \varkappa, \varkappa-1, \varkappa-2, \ldots, 0$$
$$J' + J \geq \varkappa$$
$$\Delta M = M' - M = -\varkappa, -\varkappa + 1, \ldots, \varkappa$$

Additionally, the parity of the atomic states change as

$(-1)^{\varkappa}$ in an electric multipole transition

$(-1)^{\varkappa+1}$ in a magnetic multipole transition

This rules are inviolate. In specific cases, such as in the *LS* coupling or *jj* coupling approximation, additional selection rules can be formulated, the fulfilment of which depends on to what extent the approximation used is valid.

* Electric dipole transitions, $\varkappa = 1$, $LSJM$ — $L'S'J'M'$ are allowed if the following selection rules are satisfied:

$$\Delta J = J - J' = 0, \pm 1$$
$$J + J' \geq 1$$
$$\Delta M = M - M' = 0, \pm 1$$
$$\text{odd term} \rightleftarrows \text{even term}$$

If the spin-orbit interaction is small (*LS* coupling approximation), the spin of an atom does not change in an electric dipole transition, thus

$$\Delta S = 0$$
$$\Delta L = 0, \pm 1$$
$$L + L' \geq 1$$

Because $S = 0$, allowed transitions are possible only between terms of one multiplicity. Transitions between terms of different multiplicities are forbidden (<u>intercombination</u> <u>transitions</u>). Fulfilment of this rule is evidence for the validity of the *LS* coupling approximation.

* Magnetic dipole radiation is allowed in the *LS* coupling approximation if

$$\Delta S = 0$$
$$\Delta L = 0$$
$$\Delta J = \pm 1$$

The probability of magnetic dipole radiation is α_F^2 times smaller the probabil ity of electric dipole radiation of the same frequency; $\alpha_F \rightarrow$ fine structure constant.

* Electric quadrupole radiation, $\varkappa = 2$, satisfies the selection rules

$$\Delta J = 0, \pm 1, \pm 2$$
$$J + J' \geq 2$$

The electric quadrupole transition is possible only between states of the same parity.

In the *LS* coupling approximation:

$$\Delta S = 0$$
$$\Delta L = 0, \pm 1, \pm 2$$
$$L + L' \geq 2$$

In *jj* the coupling approximation:

$$\Delta j = 0, \pm 1, \pm 2$$
$$j + j' \geq 2$$

Reference: [121]

12.4. ATOMIC POLARIZABILITY

The atomic polarizability α is defined as the induced dipole moment per unit electric field. For steady or low frequency field (indicated by "-") $\bar{\alpha}$ can be written as

$$\bar{\alpha} = 4\, a_0^3 \sum_n \frac{f_{n1}}{\left(\frac{\nu_{n1}}{Ry_\infty}\right)^2}$$

where ν_{n1}/Ry_∞ is the frequency in Rydbergs ($Ry_\infty = 3.2898 \times 10^{15}\ s^{-1}$) of lines connecting the ground level, f_{n1} denotes the corresponding oscillator strength and a_0 is the → Bohr radius.

Numerically,

$$\bar{\alpha} = \begin{cases} 6.415 \sum f_{n1}/\nu^2_{n1} & (SI) \\ 6.415 \times 10^6 \sum f_{n1}/\nu^2_{n1} & (CGS) \end{cases}$$

with

	$\bar{\alpha}$	f_{n1}	ν_{n1}
SI	m^3	-	s^{-1}
CGS	cm^3	-	s^{-1}

or

$$\bar{\alpha} = \begin{cases} 7.138 \times 10^{-17} \sum f_{n1}\, \lambda^2_{n1} & (SI) \\ 7.128 \times 10^{-23} \sum f_{n1}\, \lambda^2_{n1} & (CGS) \end{cases}$$

with

	$\bar{\alpha}$	f_{n1}	λ_{n1}
SI	m^3	-	m
CGS	cm^3	-	cm

Values of $\bar{\alpha}$ are given in the table below for some elements:

element	$\bar{\alpha}$ in $10^{-19}\ m^3$ or in $10^{-25}\ cm^3$
H	6.70
He	2.07
Li	200
Be	93
O	1.5
Ne	3.96
Na	270
Ar	16.54
K	380
Kr	24.8
Rb	500
Xe	40.4
Cs	500
Hg	52

Reference: [78] p.86

13. PLASMA DIAGNOSTICS

13.1. COMPILATION OF SOME PLASMA DIAGNOSTIC TECHNIQUES

The compilation is structured as follows:

Plasma quantity to be determined

method and/or device, range for which the method is applicable,

pecularities of the method, references.

Naturally, this compilation is not complete due both to the complexity of the methods and to the rapid development of new techniques. For some methods, formulae are given in sections 13.2. and 13.3.

1. Electron density

(1) microwave interferometry, $10^{16}\ m^{-3} < n_e < 10^{20}\ m^{-3}$, [155], [42]

(2) microwave cavity pertubation, $10^{14}\ m^{-3} < n_e < 10^{18}\ m^{-3}$, [42]

(3) microwave scattering, $10^{18}\ m^{-3} < n_e < 10^{20}\ m^{-3}$, sensitive to instabilities, [42]

(4) optical interferometry (laser interferometry, holographic interferometry, $10^{14}\ m^{-3} < n_e < 10^{18}\ m^{-3}$, [30], [59], [147], [150], [151], [155]

(5) optical spectroscopic intensities, $n_e > 10^{19}\ m^{-3}$, equilibrium plasmas, [2], [52], [103], [149]

(6) Stark broadening of spectral lines, $n_e > 10^{22}\ m^{-3}$, [57], [148]

(7) optical scattering (Laser scattering), $10^{20}\ m^{-3} < n_e < 10^{25}\ m^{-3}$[30], [59], [155]

(8) optical Faraday rotation, $n_e > 10^{22}\ m^{-3}$, suitable for $B > 1$ T, [30]

(9) Langmuir probes and other particle collectors, $10^{-2} < j_e = en_ev_e < 10^{4}\ A/m^2$ [132]-[146], [197], [198]

(10) electron and ion beam scattering, sensitive to potential fluctuations

2. Electron temperature

(1) microwave radiation intensities, $T_e \geq 0.1$ eV, stable plasmas, [42]

(2) infrared and optical intensities, $T_e \geq 1$ eV, equilibrium plasmas, [2], [152], [52]

(3) relative intensities of spectral lines, $0.5\ eV < T_e < 50\ eV$, [2], [52], [103]

(4) relative intensities of bremsstrahlung and recombination radiation, [2]

(5) X-ray intensities, $T_e \geq 6$ keV, [153]

(6) Doppler broadening of optical spectral lines, $T_e \geq 0.5$ eV, [2], [52], [103]

(7) Doppler broadening of cyclotron radiation lines, $T_e \geq 50$ eV

(8) Doppler broadening of optical Thomson scattering, $T_e \geq 5$ eV, [59]

(9) Langmuir probes and other particle collectors, $0.1\ eV < T_e < 1\ keV$, low and moderate densities, [132]-[146], [197], [198]

3. Ion density

(1) Stark broadening of spectral lines, $n_i > 10^{21}$ m^{-3}, [56], [57], [103]

(2) single and double Langmuir probes, [132]-[146], [197], [198]

(3) electron, ion, atom, or neutron beam probes, $n_i > 10^{20}$ m^{-3}

(4) Alfvén and sound wave propagation

(5) calorimetry, requires the knowledge of temperature

(6) charge-exchange neutral detection

4. Ion temperature and energy

(1) Doppler broadening of optical spectral lines, $T_i \geq 0.5$ eV, [2], [52], [103]

(2) Time-of-flight measurment gives particle or shock front velocity: sampling probes, high-speed photography

(3) calorimetry for determining the total energy and momentum [197], [198]

(4) magnetic probes (diamagnetic effects)

5. Neutral density and identification

(1) shielded ionization gauge, [156]

(2) schlieren and Mach-Zehnder photography, [30]

(3) ion or neutral atom beam scattering

(4) Rayleigh scattering of photons by bound electrons

(5) resonance absorption of photons by bound electrons, [150], [151]

(6) charge exchange detectors, fast neutrals

(7) molecular resonance spectroscopy in the RF in IR spectral region

(8) laser induced flourescence (LIF), [164], [165], [168]

(9) coherent anti-Stokes Raman scattering (CARS), [164]

(10) optogalvanic spectroscopy, [164], [168]

(11) resonance interferometry (e.g. hook method [167])

6. Drift velocity, shock velocity, rotation, and thrust

(1) Doppler shift of spectral lines, [2], [52], [103]

(2) Doppler shift of synchrotron radiation

(3) Doppler frequency of reflected microwaves, [42]

(4) time-of-flight measurments, probes, light and microwave sampling

(5) ballistics and calorimetry

(6) charge exchange detectors, fast neutrals

(7) laser interferometry and holography, [150], [151], [155]

7. Instabilities and turbulence

(1) electron and ion energy/momentum analysis

(2) high-speed photography

(3) non-thermal effects of microwave radiation, [42]

(4) microwave scattering from turbulence, [42], [59]

(5) time-resolved spectroscopy

(6) Langmuir probes, RF probes, magnetic probes and Rogowski loops

(7) external voltage-current measurments, [154]

(8) X-ray spectroscopy, high-temperature plasma, [153]

8. Constituent identification

(1) mass spectrometry, [156]

(2) optical emission spectroscopy, [2], [52]

(3) optical absorption spectroscopy, [2], [52]

(4) resonance interferometry, [150], [151]

(5) ion cyclotron resonance absorption

(6) coherent anti-Stokes Raman scattering (CARS), [164]

9. Electric fields in plasmas

(1) optogalvanic detection of Stark-shifted Rydberg states [169]

(2) measurment of Stark broadening and shift by emission or absorption spectroscopy

(3) detection of Stark shift using LIF [166] or CARS techniques

(4) probe field analyser

13.2. PROBE DIAGNOSTICS

13.2.1. GENERAL

Different classes of probe diagnostics can be distinguished in terms of the relations between L, λ_D, and λ (or λ_D/L, λ/L, and λ/λ_D) where

L : characteristic length of the probe (e.g. its radius R)

λ_D : → Debye length

λ : → mean-free-path

These classes are [132]

(A) Classical Langmuir probe

(A1) $\lambda \gg L \gg \lambda_D$: (space-charge limited) thin sheath (conventional Langmuir probe)

(A2) $\lambda \gg \lambda_D \gg L$: orbital-motion limited thick sheath

(A3) $\lambda_D \gg \lambda \gg L$: collisional thick sheath

(B) Continuum electrostatic probe

(B1) $L \gg \lambda_D \gg \lambda$: collisional thin sheath

(B2) $\lambda_D \gg L \gg \lambda$: collisional thick sheath

(B3) $L \gg \lambda \gg \lambda_D$: collisionless thin sheath (dense plasma)

Most probe theories are asymptotic to one of the three parameters λ, λ_D, and L. In the next sections, only the essentials of probe theories are given. Elaborate probe theories can be found in the literature.

References: [132]- [141]

13.2.2. THE SINGLE LANGMUIR PROBE

We consider a single probe situated in a plasma under the conditions (A1). Additionally we assume

(i) a Maxwellian distribution of electrons and ions in the undisturbed plasma

(ii) only singly ionized ions

Under these conditions the probe will have an I-U characteristic as shown in the figure.

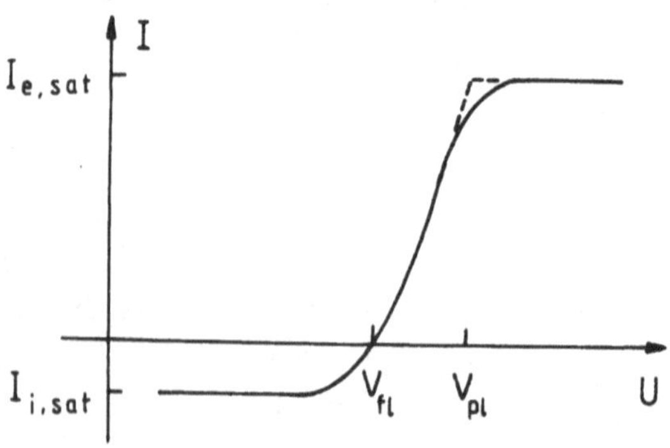

In the absence of a magnetic field, the ion saturation current at very negative probe potential can be expressed as

$$I_{i,sat} = A\, e\, n_{io}\, \overline{v}_{i\perp} \tag{1}$$

where A is the sampling area (probe surface) and n_{io} the ion density of the undisturbed plasma. According to Langmuir's approach, the mean ion velocity towards the probe surface, $\overline{v}_{i\perp}$, can be approximated by

$$\bar{v}_{i\perp} = \frac{\bar{v}_i}{4} = \frac{1}{4}\left(\frac{8\,k\,T_i}{\pi\,m_i}\right)^{1/2} \tag{2}$$

The electron current is

$$I_e = I_{e,sat}\, exp\left(-\frac{e\,U}{k\,T_e}\right) \tag{3}$$

with the electron saturation current

$$I_{e,sat} = A\,e\,n_{eo}\,\frac{1}{4}\left(\frac{8\,k\,T_e}{\pi\,m_e}\right)^{1/2} \tag{4}$$

U is the probe potential with respect to the plasma potential V_{pl}.

According to Bohm's approach [134], the transition from the undisturbed plasma to the probe should be divided into two regions:

(I) a collisionless space charge sheath adjecent the probe surface, in which large potential and density gradients and a strong electric field exist

(II) a quasi-neutral presheath with weak gradients of potential and density

The mean ion velocity towards the probe surface at the presheath-sheath interface is given by

$$\bar{v}_i = \left[\frac{2}{m_i}\left(e\,V + \frac{k\,T_i}{3}\right)\right]^{1/2} \tag{5}$$

where V is the potential at the interface. The electron (and ion) density at the interface is given by

$$n_e = n_{eo}\, exp\left(-\frac{e\,V}{k\,T_e}\right) \cong n_i \tag{6}$$

where n_{eo} denotes the density of the undisturbed plasma. The ion current density is thus

$$j_i = n_{eo}\, exp\left(-\frac{e\,V}{k\,T_e}\right)\left[\frac{2}{m_i}\left(e\,V + \frac{k\,T_i}{3}\right)\right]^{1/2} \tag{7}$$

and the ion saturation current

$$I_{i,sat} = A\,e\,n_{eo}\, exp\left(\frac{T_i}{3\,T_e} - \frac{1}{2}\right)\left(\frac{k\,T_e}{m_i}\right)^{1/2} \tag{8}$$

For $T_e > T_i$,

$$I_{i,sat} = A\ e\ n_{eo}\ 0.61 \left(\frac{k\ T_e}{m_i}\right)^{1/2} \qquad (9)$$

Note that the *electron* temperature occurs in Bohm's expressions (8) and (9) in contrast to Langmuir's result (1) and (2).

An insulated (floating) probe will be negatively charged with respect to the plasma potential due to the large ion/electron mass ratio. The total current to a floating probe is zero:

$$j_{i,sat} + j_e + j_{SE(i)} + j_{SE(e)} = 0 \qquad (10)$$

where $j_{SE(i,e)}$ denotes the current density of secondary electrons caused by the impact of primary ions and electrons on the probe surface. To satisfy eq.(10), the probe potential must attain the floating potential

$$\begin{aligned} V_{fl} &= -\frac{k\ T_e}{e}\left[1 + \ln\left(\frac{m_i}{2\pi\ m_e}\right) - 2\ \ln\left(\frac{1+\gamma_i}{1-\gamma_e}\right) - \frac{2}{3}\frac{T_i}{T_e}\right] \\ &\cong -\frac{k\ T_e}{e}\ \ln\left(\frac{m_i}{2\pi\ m_e}\right) \end{aligned} \qquad (11)$$

To determine of the electron temperature, one may use the slope of the electron current (3)

$$T_e = \frac{e}{k}\ \frac{1}{d\left[\ln\left(I_e/I_{e,sat}\right)\right] / dU} = \frac{11605}{d\left[\ln\left(I_e/I_{e,sat}\right)\right] / dU} \qquad (12)$$

	T_e	U
SI	K	V
CGS	K	V

The electron density can be determined from the electron saturation current (4):

$$n_{eo} = \frac{4\ I_{e,sat}}{e\ A}\left(\frac{\pi\ m_e}{8\ k\ T_e}\right)^{1/2} \qquad (13)$$

$$n_{eo} = \begin{cases} 4.03 \times 10^{15} \; I_{e,sat}/(A \; T_e^{1/2}) & (SI) \\ 4.03 \times 10^{13} \; I_{e,sat}/(A \; T_e^{1/2}) & (CGS) \end{cases}$$

	n_e	I_e	A	T_e
SI	m^{-3}	A	m^2	K
CGS	cm^{-3}	A	cm^2	K

Under the assumptions of (A1), the space charge sheath is very thin and the above formulae are valid for any probe geometry.

In the orbital-motion limited case of a thick sheath (conditions (A2)), the following relations hold (r_s is the border of the space charge sheath):

(i) for a flat probe

$$j = \begin{cases} \dfrac{e\, n(r_s)}{4} \left(\dfrac{8\, kT}{\pi\, m}\right)^{1/2} \exp\left(\dfrac{e\, U}{kT}\right) & \text{for } U \le V_{pl} \equiv 0 \\ \dfrac{e\, n(r_s)}{4} \left(\dfrac{8\, kT}{\pi\, m}\right)^{1/2} & \text{for } U \ge V_{pl} \equiv 0 \end{cases} \tag{14}$$

(ii) for a cylindrical probe

$$j = \begin{cases} \dfrac{en(r_s)}{4} \left(\dfrac{8\, kT}{\pi\, m}\right)^{1/2} \exp\left(\dfrac{e\, U}{kT}\right) & \text{for } U \ge V_{pl} \equiv 0 \\ \dfrac{en(r_s)}{4} \left(\dfrac{8\, kT}{\pi\, m}\right)^{1/2} \left\{ \dfrac{r_s}{R}\, \mathrm{erf}\left[\left(\dfrac{R^2}{r_s^2 - R^2}\, \dfrac{eU}{kT}\right)^{1/2}\right] + \exp\left(\dfrac{eU}{kT}\right) \left[1 - \mathrm{erf}\left[\left(\dfrac{R^2}{r_s^2 - R^2}\, \dfrac{eU}{kT}\right)^{1/2}\right]\right] \right\} & \\ \qquad \text{for } U \le V_{pl} \equiv 0 & \end{cases} \tag{15}$$

(iii) for a spherical probe

$$j = \begin{cases} \dfrac{en(r_s)}{4}\left(\dfrac{8\ kT}{\pi\ m}\right)^{1/2} \exp\left(\dfrac{eU}{kT}\right) & \text{for } U \geq V_{pl} \equiv 0 \\ \\ \dfrac{en(r_s)}{4}\left(\dfrac{8\ kT}{\pi\ m}\right)^{1/2} \dfrac{r_s^2}{R^2}\left\{1 - \dfrac{r_s^2-R^2}{r_s^2} \exp\left[-\dfrac{R^2}{r_s^2-R^2}\,\dfrac{eU}{kT}\right]\right\} & \text{for } U \leq V_{pl} \equiv 0 \end{cases} \tag{16}$$

$n(r_s)$ is the density (of electrons or ions) at the sheath edge; it is frequently used as an approximation for the density in the undisturbed plasma. For sufficiently high voltages, $eU \gg kT$, the formulae can be simplified as follows for a cylindrical probe

$$j = \frac{e\ n(r_s)}{4}\left(\frac{8\ kT}{\pi\ m}\right)^{1/2}\left(1 + \frac{eU}{kT}\right)^{1/2} \tag{17}$$

for a spherical probe

$$j = \frac{e\ n(r_s)}{4}\left(\frac{8\ kT}{\pi\ m}\right)^{1/2}\left(1 + \frac{eU}{kT}\right) \tag{18}$$

For cylindrical probes, the electron density can be determined from

$$n_e = \frac{\pi\ I_e}{e\ A}\left(\frac{m_e}{2\ kT_e}\right)^{1/2}\left(1 + \frac{eU}{kT_e}\right)^{-1/2} \tag{19}$$

$$= \begin{cases} 3.56 \times 10^{15}\ \dfrac{I_e}{A}\left(T_e + 11605\ U\right)^{-1/2} & \textit{(SI)} \\ \\ 3.56 \times 10^{13}\ \dfrac{I_e}{A}\left(T_e + 11605\ U\right)^{-1/2} & \textit{(CGS)} \end{cases} \tag{20}$$

and for spherical probes from

$$n_e = \frac{4\ I_e}{e\ A}\left(\frac{\pi\ m_e}{8\ kT_e}\right)^{1/2}\left(1 + \frac{eU}{kT_e}\right)^{-1} \tag{21}$$

$$n_e = \begin{cases} 4.03 \times 10^{15} \dfrac{I_e}{A} T_e^{1/2} \left(T_e + 11605\ U\right)^{-1} & (SI) \\ 4.03 \times 10^{13} \dfrac{I_e}{A} T_e^{1/2} \left(T_e + 11605\ U\right)^{-1} & (CGS) \end{cases} \quad (22)$$

	n_e	T_e	U	I_e	A
SI	m^{-3}	K	V	A	m^2
CGS	cm^{-3}	K	V	A	cm^2

Probe characteristics under conditions (A1) and (A2) are shown in the figure below for flat (a), cylindrical (b), and spherical (c) probe geometry ([136] pp.56, 57)

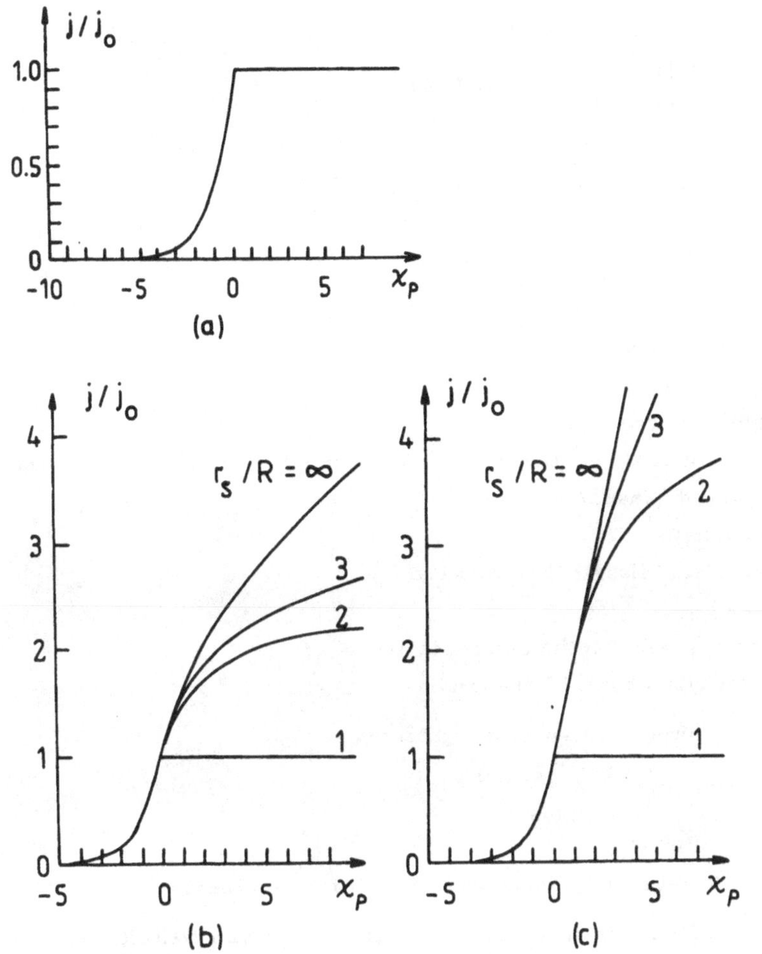

In these figures the current density to the probe, j, is normalized by

$$j_0 = \frac{n_{e0}\, e}{4} \left(\frac{8\, kT}{\pi\, m} \right)^{1/2}$$

whilst the probe voltage is normalized by kT, i.e.

$$\chi_p = \frac{eU}{kT}$$

References: [133]-[139]

13.2.3. SINGLE PROBE UNDER CONTINUUM CONDITIONS

Current collection by a probe under continuum conditions depends on a large number of parameters,

$$\hat{j} = \hat{j}\left(\left(\frac{\lambda_D}{L} \right)^2,\ \chi_p,\ \chi_p \tau,\ ReSc_i,\ \mathfrak{D},\ \mathfrak{D}_e,\ T_0/T_w \right)$$

where $\hat{j}$ denotes the dimensionless probe current

$$\hat{j} = \frac{I}{A\, n_{e0}\, Z\, e\, v}$$

with

- I : (dimensional) probe current
- A : probe surface area
- e : electronic charge
- n_{e0} : electron density at the reference point 0 (for example, in the undisturbed plasma)
- Z : charge number
- v : characteristic plasma flow velocity

and

- λ_D : → Debye length (at the reference point)
- L : characteristic length of the probe
- $\chi_p = \frac{eU}{kT_{e0}}$: dimensionless probe potential
- $\tau = T_{e0}/T_{i0}$
- $ReSc_i = \frac{u_0\, L}{D_{i0}}$: ratio of convective to diffusive transport (electric Reynolds number), u_0 is the convection velocity at the reference point, D_{i0} the ion diffusion coefficient

$\mathfrak{D}_e, \mathfrak{D}$: $\rightarrow$ Damköhler number for thermal and chemical equilibration respectively

T_0/T_w : ratio of the temperature at the reference point to the wall (probe) temperature

Naturally, no general expression exists for the I-U characteristic of a probe under continuum conditions. There are a small number of analytic approximations which are valid in specific cases.

(1) Small convection limit, $ReSc_i \longrightarrow 0$

(1.1) Frozen chemistry, constant property plasma,

thin sheath limit, $|\chi_p| \left(\lambda_D/R\right)^2 \ll 1$

The solution of the governing equations (ion conservation equation, electron conservation equation, Poisson's equation) can be given analytically in the limit $\lambda_D/R \ll 1$ and $|\chi_p| \longrightarrow \infty$ [142]:

$$I = 4\pi\ R\ n_{e0}\ e\ D_i \left(-\ \chi_p \frac{\lambda_D}{R}\right)^{1/2} (1+\tau)$$

If the probe is only slightly negative with respect to the plasma potential, the expression becomes

$$I = 4\pi\ R\ n_{e0}\ e\ D_e \left\{\left(1 - \frac{D_i}{D_e}\right) - \frac{3}{2}\frac{eU}{kT_e}\left(1 + \tau\frac{D_i}{D_e}\right)\left[\ln\left(\lambda_D/R\right)\right]^{-1} + \dots\right\}$$

(1.2) Frozen chemistry, constant property plasma,

thick sheath limit, $|\chi_p| \left(\lambda_D/R\right)^2 \gg 1$, but $\lambda_D/R \le O(1)$

For a *spherical probe* of radius R, the electron or ion current is given by [143]

$$I_{e,i} = 4\pi\ R\ \hat{R}_s\ n_{e0}\ \mu_{e,i}\ k\left(T_i + T_e\right)$$

$$I_{e,i} = \begin{cases} 3.08 \times 10^{-18}\ Z\ \lambda_{en}\ R\ \hat{R}_s\ n_{eo}\left(T_i + T_e\right) / \left(m_{e,i}\ T_{e,i}\right)^{1/2} & (SI) \\ 9.73 \times 10^{-15}\ Z\ \lambda_{en}\ R\ \hat{R}_s\ n_{e0}\left(T_i + T_e\right) / \left(m_{e,i}\ T_{e,i}\right)^{1/2} & (CGS) \end{cases}$$

where

$$\hat{R}_s = 1 + \begin{cases} 0.820 \left\{\left[n_{e0} \left(T_i + T_e\right)\right]^{-1/2} |U| \;/\; R\right\}^{0.535} & (SI) \\ 388 \left\{\left[n_{e0} \left(T_i + T_e\right)\right]^{-1/2} |U| \;/\; R\right\}^{0.535} & (CGS) \end{cases}$$

and

$$\mu_{e,i} = \frac{e\,D_{e,i}}{k\,T_{e,i}}$$

	I	R	$\hat{R}_s$	n_{e0}	λ_{en}	$T_{e,i}$	U	$\mu_{e,i}$	$D_{e,i}$	$m_{e,i}$
SI	A	m	-	m^{-3}	m	K	V	m^2/Vs	m^2/s	kg
CGS	A	cm	-	cm^{-3}	cm	K	V	cm^2/Vs	cm^2/s	g

For a *cylindrical probe* of length L_c,

$$I_{e,i} = 2\pi\, L_c\, \hat{R}_s\, n_{e0}\, \mu_{e,i}\, k\left(T_i + T_e\right) \;/\; \ln\left(\pi\, L_c/4\, R\, \hat{R}_s\right) =$$

$$= \begin{cases} 1.54 \times 10^{-18}\, Z\, \lambda_{en}\, L_c\, n_{eo} \left(T_i + T_e\right) / \left(m_{e,i}\, T_{e,i}\right)^{1/2} \ln\left(\pi\, L_c/4\, R\, \hat{R}_s\right) & (SI) \\ 4.87 \times 10^{-15}\, Z\, \lambda_{en}\, L_c\, n_{e0} \left(T_i + T_e\right) / \left(m_{e,i}\, T_{e,i}\right)^{1/2} \ln\left(\pi\, L_c/4\, R\, \hat{R}_s\right) & (CGS) \end{cases}$$

where

$$\hat{R}_s = 1 + \begin{cases} 2.899 \left\{\left[n_{e0} \left(T_i + T_e\right)\right]^{-1/2} |U| \;/\; R\right\}^{0.62} & (SI) \\ 3650 \left\{\left[n_{e0} \left(T_i + T_e\right)\right]^{-1/2} |U| \;/\; R\right\}^{0.62} & (CGS) \end{cases}$$

(2) Large convection limit, $ReSc_i \gg 1$

There are two characteristic lengths:

1) one in the flow direction, L (e.g. probe radius R)

2) the second in the direction normal to the flow, boundary-layer thickness δ

(2.1.) Thin sheath limit

For a vanishingly thin sheath, the dimensionless ion-saturation current density is given by [144]

$$\hat{j} = \frac{j_{i,sat}\,(s/l)^{1/2}}{e\, n_{e\delta}\, u_\delta\, r^{\varepsilon}\, \mu_\delta} = \frac{0.47}{2^{1/2}} \left(\frac{2}{Sc_i} \right)^{2/3}$$

when $T_e = T_i$ and the inequality

$$\hat{a} = \frac{2\, l\, s}{(\rho_\delta\, u_\delta)^2\, r^{2\varepsilon}\, \lambda_D^2} \sim \frac{\delta}{\lambda_{D,\delta}} \geq 10^4$$

is satisfied. In these formulae

$$s = \int_0^x \rho_\delta\, \mu_\delta\, u_\delta\, r^{2\varepsilon}\, dx \qquad \text{and} \qquad l = \left(\frac{\rho_w\, \mu_w}{\rho^*\, \mu^*} \right)^{0.2}$$

where ρ^*, μ^* denote respectively the gas density and viscosity at the maximum temperature point in the boundary layer, ε is zero and unity for two-dimensional and axisymmetric bodies respectively, and r is the distance between the surface and the axis of symmetry. The subscript δ refers to conditions at the boundary layer edge, ρ_δ is normalized. Chung's solution is the diffusion-controlled limit for the flux of any chemically frozen species i diffusing through the boundary layer to the wall.

Explicit expressions can be obtained using the appropriate definitions of s. They are

$$\hat{j} = \frac{j_{i,sat}}{e\, n_{e\delta}\, u_\delta} \left(\frac{Re_x}{l} \right)^{1/2} \qquad \text{flat plate}$$

$$\hat{j} = \frac{j_{i,sat}}{e\, n_{e\delta}\, u_\delta} \left(\frac{Re_x}{3\, l} \right)^{1/2} \qquad \text{cone}$$

where

$$Re_x = \frac{\rho_\delta\, u_\delta\, x}{\mu_\delta}$$

(2.2.) Thick sheath limit

No general solutions are available. Various flat-plate experimental data are described by the solution of Russo and Touryan [145],

$$\hat{j} = \frac{j_{i,sat}}{e\, n_{e\delta}\, u_\delta} \left(Re_{x,\delta}\right)^{1/2} = 1.8 \left(\frac{\lambda_{D,\delta}}{R}\right)^{0.4} \left(-\frac{\chi_p}{Sc_i}\right)^{0.5}$$

where $Sc_i = \frac{\mu}{\rho D_i}$ is the ion → Schmidt number, and $Re_{x,\delta}$ is the → Reynolds number based on the distance from the plate leading edge to the probe location. Using the known general behaviour of boundary layers, the last formula can be rewritten as follows ([132] p.83)

$$\hat{j} = \frac{j_{i,sat}\, (s/l)^{1/2}}{e\, n_{e\delta}\, u_\delta\, r^{\varepsilon}\, \mu_\delta} = 1.8 \left(\frac{\lambda_{D,\delta}}{R}\right)^{0.4} \left(-\frac{\chi_p}{Sc_i}\right)^{0.5}$$

Reference: [132]

13.2.4. SINGLE PROBE IN THE CASE OF A COLLISIONLESS THIN SHEATH (DENSE PLASMA)

The ordering of characteristic lengths $L \gg \lambda \gg \lambda_D$ decribes the situation in which the charged particle motion in the quasi-neutral plasma is governed by continuum processes (drift, diffusion). In contrast, in the space-charge sheath the charged particle motion is collisionless.

The ion density at the sheath edge can be determined from Bienkowski's expression [146]

$$\left(n_i\right)_s = \hat{\tau}_s \frac{j_{i,wall}\, (m_i/2\, kT_e)^{1/2}}{(1+\hat{\tau}_s)\, A_0(\hat{\tau}_s)}$$

where $j_{i,wall}$ is the measured ion flux density to a wall at floating potential, $\hat{\tau}_s = \left(T_e/T_i\right)_s$, and $A_0(\hat{\tau}_s)$ is a function of order unity. This expression is a modified form of Bohm's sheath criterion.

The boundary-layer ion density, $\left(n_i\right)_\delta$, can be determined by

$$j_{i,wall} = \left(n_i\right)_\delta \frac{e\, Re_x^{1/2}}{x} D_a \left(\frac{\rho}{\rho_\delta}\right)^2 \frac{\mathcal{M}'(\eta_\lambda)}{2\, l^{1/2}}$$

where

$$\eta_\lambda = \frac{r^\varepsilon u_\delta}{(2 s l)^{1/2}} \int_0^\lambda \rho \, dy , \qquad \mathcal{M}' = \frac{d\mathcal{M}}{d\eta} , \qquad \mathcal{M} = C_i / C_{i,\delta} , \qquad D_a = D_i (1+\hat{\tau})$$

C_i is the ion mass fraction, and the values Re_x, l, s, r, ε, u_δ, and D_i have been defined in the preceding section.

References: [132], [146]

13.2.5. THE DOUBLE LANGMUIR PROBE

Two single Langmuir probes (for single Langmuir probe theory see section 13.2.2.) are biased with respect to one another with a voltage

$$U_d = U_1 - U_2 \tag{1}$$

but are electrically isolated from ground and from any reference electrode. The double probe system thus floats with the plasma. Therefore

$$I_d = I_1 = I_2 \tag{2}$$

For small or moderate probe voltages, $U_d < V_{pl} \equiv 0$ and

$$A_1 j_{e0} \exp \chi_{p,1} + A_2 j_{e0} \exp \chi_{p,2} = A_1 j_{i,1} + A_2 j_{i,2} \tag{3}$$

where $\chi_{p,1/2} = eU_{1/2}/ kT$ and all other symbols are defined in section 13.2.2. The ion current density, j_i, is independent of probe voltage and sheath thickness only for flat probes in a quiescent plasma,

$$j_i = j_{i0} = K \, n(r_s) \, e \left(\frac{kT_e}{m_i} \right)^{1/2} \qquad 0.5 \le K \le 2 \tag{4}$$

In this case, the probe current is given by

$$I_d = A_2 \, j_i \, \frac{\exp \chi_d - 1}{\exp \chi_d + 1/\hat{A}} = A_1 \, j_i \, \frac{\exp \chi_d - 1}{\exp \chi_d + 1/\hat{A}} \, \hat{A}^{-1} \tag{5}$$

where $\chi_d = eU_d/kT_e$ and $\hat{A} = A_1/A_2$ $(A_1 \le A_2)$. Some double probe I_d-U_d characteristics are shown in the following figure.

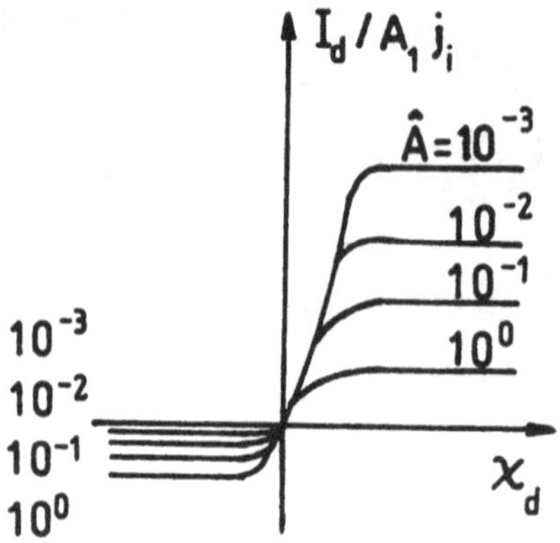

The characteristic is symmetric about zero only for equal probe area, $A_1 = A_2$. In the latter case, the characteristic can be simplified as follows

$$I_d = A\ j_i \frac{\exp \chi_d - 1}{\exp \chi_d + 1} = A\ j_i\ \operatorname{tgh}\left(\frac{\chi_d}{2}\right) \tag{6}$$

giving a useful form for experimental work.

If the probe voltage is sufficiently large, $\chi_d \gg \ln\left(A_2/A_1\right)$, the double probe current saturtes

$$I_{d,sat} = A_2\ j_i \tag{7}$$

Note that $I_{d,sat}$ is governed by the *ion* current density.

The electron temperature is determined from the slope of the characteristic (5) and the saturation current (7):

$$T_e = \frac{e}{k} \frac{\hat{A}}{1+\hat{A}} \frac{I_{d,sat}}{\left(\frac{dI_d}{dU_d}\right)_{U_d=0}} \tag{8a}$$

or, for a symmetric double probe ($\hat{A} = 1$),

$$T_e = \frac{e}{2\ k} \frac{I_{d,sat}}{\left(\frac{dI_d}{dU_d}\right)_{U_d=0}} \tag{8b}$$

The ion density is determined simply from the saturation current, (7) and (4):

$$n_i = \frac{I_{d,sat}}{A_2 \ K \ e} \left(\frac{m_i}{kT_e} \right)^{1/2} \qquad (9)$$

References: [135]-[139]

13.2.6. PROBE DIAGNOSTICS OF MAGNETIZED PLASMAS

A relatively simple analysis of particle and energy fluxes applies only to the extremes B = 0 or B very large. To avoid finite Larmor radius effects, the probe's effective area projected on the plane perpendicular to B, $A_\perp$, must satisfy the relation (strong magnetic field approximation)

$$A_\perp^{1/2} \gg r_{L,i}$$

where $r_{L,i}$ is the ion → Larmor radius.

In a non-magnetic plasma the density depression induced by probe insertion is nearly spherically symmetrical, whereas in a magnetized plasma the depression is localized along the magnetic flux tube subtended by the probe. A transverse density gradient is established between this flux tube and adjacent, relatively undisturbed flux tubes. This causes cross-field diffusion into the probe's flux tube.

The ion collection length, $L_{\parallel_i}$, is defined as the length over which the cross-field density gradient if of order n/d, where n is the plasma density and d = $A_\perp^{1/2}$. From particle balance we obtain

$$L_{\parallel_i} = \frac{v_{s,i} \ A_\perp}{8 \ D_\perp}$$

with the ion acoustic speed

$$v_{s,i} = \left(\frac{k \left(T_e + \varkappa T_i \right)}{m_i} \right)^{1/2}$$

where $\varkappa$ = 1, 5/3 for the respectively isothermal or adiabatic ion flow. $D_\perp$ is is cross-field diffusion coefficient. Under tokamak edge plasma conditions, $L_{\parallel_i}$ is in the order of a few meters or even more.

This implies:

* The probe does not provide point measurments of particle or heat flux but provides values averaged over length $L_{\parallel_i}$.
* The ion flow along the flux tube may be collisional.

Similarly, an electron collection length can be defined as

$$L_{\parallel_e} = \frac{\bar{v}_e A_\perp}{16 D_\perp}$$

where

$$\bar{v}_e = \left(\frac{8 kT_e}{\pi m_e} \right)^{1/2}$$

is the mean velocity of a → Maxwellian velocity distribution of electrons. Under tokamak edge plasma conditions, $L_{\parallel_e}$ is in the order of some hundred meters.

This implies:

* The probe measures electron quantities averaged over length $L_{\parallel_e}$.
* The electron flux to the probe is more likely collisional as the ion flux.

The electron saturation current is reduced to

$$j_{e,sat} = \left(\frac{r}{1 + r} \right) \frac{1}{4} n_{eo} \, e \, \bar{v}_e$$

where n_{eo} is the electron density in the undisturbed plasma and

$$r = \frac{16}{\pi} \left(1 + \frac{D_i}{T_e} \right) \left(\frac{D_\perp}{D_\parallel} \right)^{1/2} \frac{\lambda_{ei}}{d}$$

The parallel-field electron diffusion coefficient can be expressed as

$$D_\parallel = \frac{1}{4} \lambda_{ei} \, \bar{v}_e$$

where λ_{ei} is the electron-ion → mean-free-path.

The total current density to the probe is

$$j_{tot} = j_{i,sat} - \frac{1}{4} n_e^{(s)} \, e \, \bar{v}_e \left(1 - \gamma_{SEE} \right) \exp\left(\frac{eU}{kT_e} \right)$$

where $j_{i,sat}$ is the ion saturation current, $n_e^{(s)}$ denotes the electron density at the plasma/sheath interface, U is the probe voltage with respect to the local plasma, γ_{SEE} is the → secondary electron emission yield. Plotting experimental values for $\ln\left(j_{tot} - j_{i,sat} \right)$ vs U, the electron temperature can be deduced from the slope of this plot.

From the ion saturation current

$$j_{i,sat} = e\, n_e^{(s)}\, v_{s,i}$$

($= 0.5\, e\, n_{e0}\, v_{s,i}$ for the collisionless pre-sheath) the local plasma density, $n_e^{(s)}$, can be deduced if T_e has already been measured and if $T_e = T_i$, or a value for $v_{s,i}$ has been determined with an alternative method.

Reference: [198]

13.3. INTERFEROMETRY

The phase shift experienced by an electromagnetic beam in passing through a gas (plasma) with respect to a reference beam following the same path length in free space is usually termed experimentally as a "fringe" shift. The experimental fringe shift δ may be written in terms of the actual phase shift $\Delta\varphi$ as:

$$\delta = \frac{\Delta\varphi}{2\pi} = \frac{1}{\lambda} \int_{z_1}^{z_2} \left[\eta(x,y,z) - \eta_0\right] dz \qquad (1)$$

where

λ : wavelength of the incident electromagnetic beam

$\eta(x,y,z)$, η_0 : refractive index at the point (x,y,z) and of the undisturbed enviroment respectively

$l = z_2 - z_1$: geometrical thickness of the plasma layer under investigation

Contributions to the refractive index originating from the different species α will be superimposed

$$\eta - 1 = \sum_\alpha \left(\eta - 1\right)_\alpha = \sum_\alpha c_\alpha n_\alpha \qquad (2)$$

where n_α is the density of particles α.

The electron contribution is usually given by

$$\left(\eta - 1\right)_e = -\frac{1}{2}\left(\frac{\omega_{pl,e}}{\omega}\right)^2 = -K_\varepsilon \frac{e^2 \lambda^2 n_e}{2\pi m_e c^2} \qquad (3)$$

$$= \begin{cases} -4.485 \times 10^{-16} \lambda^2 n_e & (SI) \\ -4.485 \times 10^{-14} \lambda^2 n_e & (CGS) \end{cases}$$

with

	λ	n_e
SI	m	m^{-3}
CGS	cm	cm^{-3}

and $K_\varepsilon = \begin{cases} 1/(4\pi\,\varepsilon_0) & (SI) \\ 1 & (CGS) \end{cases}$

$\omega_{pl,e}$ is the → electron plasma frequency and $\omega = 2\pi\, c/\lambda$ the angular frequency of the incident radiation. Eq.(3) holds in the absence of an external magnetic field and if $\nu \ll \omega$, where ν denotes the collision frequency. As illustrated in the figure below, the sensitivity of interferometric determination of the electron density increases with wavelength λ. In the figure, l is the plasma length under investigation and the parameter $\Delta\varphi/2\pi$ is the magnitude of the fringe shift.

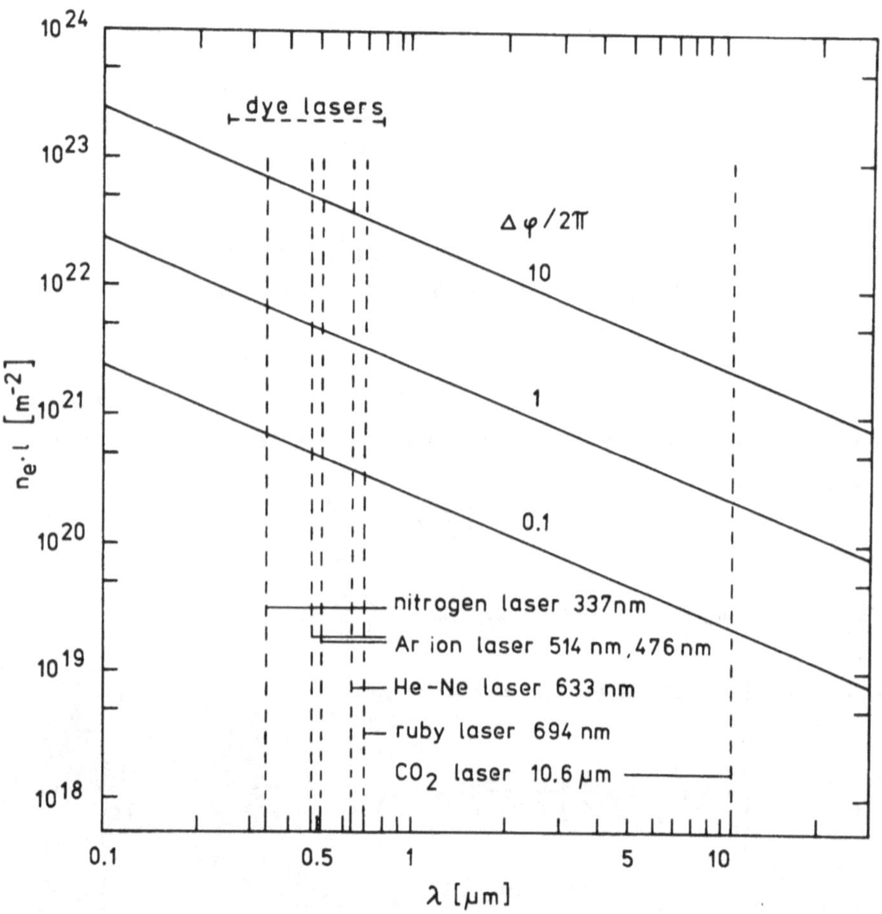

The contribution of heavy particles (neutrals, ions) can be written as

$$\left(\eta - 1\right)_h = K_\varepsilon \frac{2\pi e^2}{m_e} \sum_{\substack{u,l \\ u \neq l}} \frac{f_{lu} n_l}{\omega_{lu}^2 - \omega^2} = K_\varepsilon \frac{2\pi e^2}{m_e} \sum_{\substack{u,l \\ u \neq l}} \frac{f_{lu} \lambda^2 \lambda_{lu}^2 n_l}{\lambda^2 - \lambda_{lu}^2} \qquad (4)$$

where n_l is the particle density in the state *l*, f_{lu} denotes the → absorption oscillator strength of the optical transition $l \longleftrightarrow u$. This formula is only valid when the incident radiation is clearly outside of the half width of any absorption line. As a rule, the strongest absorption lines (resonance lines) are in the UV-region, where $\lambda \gg \lambda_{lu}$ and by series expansion of (4) the often used Cauchy formula is obtained

$$\left(\eta - 1\right)_h = \left[A + B/\lambda^2\right] \frac{n_h}{n_L} \qquad (5)$$

n_h denotes the heavy particle density and n_L is the → Loschmidt constant (= 2.6868 x 10^{25} m^3). A and B are gas specific constants (see table below).

gas	A in 10^{-6}	B in 10^{-18} m^{-2} with λ in m, or in 10^{-14} cm^{-2} with λ in cm
He	34.8	0.08
Ne	66.6	0.16
Ar	279.2	1.56
Kr	418.9	9.92
Xe	682.3	6.92
Hg	878	19.9
air	287.1	1.63
N_2	290.6	2.24
H_2	135.8	1.02
O_2	266.3	2.24
CO	327	2.65
CO_2	439	2.81
NO	289	2.02
NH_3	370	4.44

References: [30] p.9, [52] p.228, [76], [77], [78] p.87, [79]

In the visible region, $A \gg B/\lambda^2$ and so $(\eta - 1)_h$ is practically independent of λ.

If the incident radiation is within the half width, $\Delta\lambda$, of an absorption line, eq.(4) must be replaced by

$$\left(\eta - 1\right)_h = \frac{r_e}{4\pi} \frac{\lambda_{lu}^3 n_l f_{lu} (\lambda - \lambda_{lu})}{(\lambda - \lambda_{lu})^2 + (\Delta\lambda/2)^2} \tag{6}$$

where $r_e = K_\varepsilon \dfrac{e^2}{4\pi m_e c^2}$ is the → classical electron radius. For a very narrow line, when $\left(\lambda - \lambda_{lu}\right)^2 \gg \left(\Delta\lambda/2\right)^2$, the Sellmeier formula holds:

$$\left(\eta - 1\right)_h = \frac{r_e}{4\pi} \frac{\lambda_{lu}^3 n_l f_{lu}}{\lambda - \lambda_{lu}} \tag{7}$$

References: [79], [80]

14. ELECTRODE PHENOMENA

14.1. THERMIONIC EMISSION OF ELECTRONS

The current density of electrons emitted from a hot metallic surface is given by the Richardson formula (Richardson-Dushman formula)

$$j_T^e(T) = D\, A_0\, T^2 \exp\left(-\frac{\Phi_E}{kT}\right)$$

where

$$A_0 = \frac{4\pi\, m_e\, e\, k^2}{h^3} = 1.20 \times 10^6\ \mathrm{A/\ m^2\ K^2} = 120\ \mathrm{A/\ cm^2\ K^2}$$

D correction factor depending on the emittting material

Φ_E effective work function

The reduction of the work function Φ by an electric field is the Schottky effect

$$\Phi_E = \Phi - \left(K_\varepsilon\, e^3\, E\right)^{1/2}$$

with $K_\varepsilon = \begin{cases} 1/4\pi\,\varepsilon_0 & (SI) \\ 1 & (CGS) \end{cases}$, E is the electric field strength perpendicular to the emitter surface. A good approximation is, $D \cong 0.5$. Often, the so-called Richardson constant, $A = D\, A_0$ is used.

For several important electrode materials, values of A and Φ are given in the following table

material	A in 10^6 A/m^2 K^2 or 10^2 A/cm^2 K^2	Φ in 10^{-19} J or 10^{-12} erg	or in eV
Re	7.20	7.59	4.74
C	0.60	7.26	4.53
Pt	0.32	8.52	5.32
Fe	0.26	7.18	4.48
Ta	0.37	6.51	4.09
Zr	3.30	6.60	4.12
Hf	0.145	5.66	3.53
Th	0.70	5.41	3.38
Nb	0.57	6.34	3.96
Mo	1.07	6.87	4.29
Cu	0.60	7.18	4.48

material	A in 10^6 A/m^2 K^2 or 10^2 A/cm^2 K^2	Φ in 10^{-19} J or 10^{-12} erg	or in eV
BaO	0.001	1.60 - 2.40	1.0 - 1.5
BaO+SrO	0.01-0.00001	1.65 - 1.92	1.03-1.20
LaB_6	0.40	3.84	2.4
ThO_2	0.50	4.17	2.6
Y_2O_3	2.09	3.20	2.0
Gd_2O_3	2.74	3.52	2.2
ZrC	1.34	6.09	3.8
Hg	0.60	7.26	4.53
Au	0.60	7.55	4.71
W	0.75	7.29	4.55
film cathodes			
W-Th	0.03	4.21	2.63
W-Ba	1.00	4.26	2.66
W-Ba-Al	0.15	3.20	2.0
W-Cs	0.0326	2.21	1.38
W-La	0.08	4.36	2.72
Ta-Th	0.01	4.04	2.52

References: [91], [92] pp.20,21, [93] p.354, [94]

Thickness of and voltage drop across the space charge zone:

Within the framework of a simplified space-charge model, the → Poisson equation may be solved analytically on the basis of the following **assumptions**:

(i) the thickness of the space-charge zone is much smaller than the electron and ion → mean-free-path

(ii) both electrons emitted from the cathode surface and ions entering the space-charge zone have vanishingly small initial velocities

(iii) the influence of the electrons emitted and of the reverse electron flux on the space-charge zone is negligibly small

The thickness of the space-charge zone is then given by

$$d \cong \frac{2}{3}\,\varepsilon_0^{1/2}\left(\frac{2\,e}{m_i}\right)^{1/4} U_c^{3/4}\; j_i^{-1/2} \qquad (SI)$$

where U_c is the cathode voltage drop and j_i denotes the current density of ions arriving at the cathode surface. The electric potential difference with respect to the cathode potential can be expressed by

$$U(x) \cong U_c\left[1 - \left(1 - \frac{x}{d}\right)^{4/3}\right] \qquad \text{for } x \le 0 \le d$$

and the electric field strength at the cathode surface is consequently (Mackeown formula)

$$E_0 = \left.\frac{\partial U}{\partial x}\right|_{x=0} \cong \frac{4}{3}\frac{U_c}{d}$$

Because assumption (iii) is not well justified, these formulae give only approximative values.

The space-charge limited current between two parallel electrodes of area A is given by the Child-Langmuir law [189]

$$I = \frac{4}{9}\,\varepsilon_0 \left(\frac{2\,e}{m_e}\right)^{1/2} \frac{U^{3/2}}{d^2}\,A = 2.334 \times 10^{-6}\,\frac{U^{3/2}}{d^2}\,A$$

where the gap voltage U is in V/m, the gap distance d in m, and the area A in m^2. This formula applies only if the work function of the cathode is negligible.

References: [92] p.28, [95]

14.2. FIELD EMISSION OF ELECTRONS

The current density of electrons emitted from a metal by an intense electric field is given by the Fowler-Nordheim formula

$$j_F^e(E) = \frac{e^3\,E^2}{8\pi\,h\,\Phi\,t^2(y)}\,\exp\left[-\frac{8\pi\left(2\,m_e\right)^{1/2}\Phi^{3/2}}{3\,e\,h\,E}\,v(y)\right]$$

where

E : electric field strength normal to the emitter surface

Φ : work function

$$y = \frac{\left(\dfrac{e^3\,E}{4\pi\,\varepsilon_0}\right)^{1/2}}{\Phi} = 3.795 \times 10^{-5}\,\frac{E^{1/2}}{\Phi}$$

with E in V/m and Φ in eV.

t(y) and v(y) are elliptical functions given in the following table

y	t(y)	v(y)	y	t(y)	v(y)
0.00	1.000	1.000	0.75	1.076	0.383
0.05	1.001	0.995	0.80	1.083	0.312
0.10	1.004	0.982	0.85	1.090	0.238
0.15	1.007	0.962	0.90	1.097	0.161
0.20	1.011	0.937	0.95	1.104	0.082
0.25	1.016	0.907	1.00	1.110	0.000
0.30	1.021	0.872	1.1	1.125	-0.172
0.35	1.026	0.832	1.2	1.139	-0.353
0.40	1.032	0.789	1.3	1.153	-0.545
0.45	1.038	0.741	1.4	1.167	-0.745
0.50	1.044	0.690	1.5	1.180	-0.954
0.55	1.050	0.635	2.0	1.249	-2.122
0.60	1.057	0.577	3.0	1.380	-5.213
0.65	1.063	0.515	4.0	1.501	-8.458
0.70	1.070	0.450	5.0	1.615	-12.438

An often used form of the Fowler-Nordheim formula is

$$j_F^e(E) = 1.54 \times 10^{-6} \frac{E^2}{\Phi\, t^2(y)} \exp\left[- \frac{6.83 \times 10^9\, \Phi^{3/2}\, v(y)}{E}\right]$$

with j in A/m^2, E in V/m and Φ in eV. If j_e is expressed in A/cm^2, E in V/cm and Φ in eV, the factor 10^9 should be replaced by 10^7.

The dependence of j_F^e on the functions t(y) and v(y) and the numerical values of these functions, indicate that t(y) has only a marginal effect, whereas v(y) strongly influences the current density. Thus, the function v(y) requires careful treatment. Hantzsche [123] has given an approximate analytic expression

$$v(y) \cong \begin{cases} 1 - y^2 \left[0.3750\, \ln(1/y) + 0.9673\right] - 0.0327\, y^4 & \text{for } y \leq 1 \\ -1.3110\, y^{3/2} + 0.8986\, y^{1/2} + 0.4936\, y^{-1/2} - 0.0812\, y^{-3/2} & \text{for } y \geq 1 \end{cases}$$

References: [96], [97] pp.17-23, [98] p.176, [123]

14.3. THERMOFIELD EMISSION OF ELECTRONS

If both the temperature of the emitter and the intense electric field at the emitter surface are essential for the electron emission, the electron current density is given by the general expression

$$j^{e}_{TF}(T,E) = \frac{4\pi\, m_e\, kT}{h^3} \left\{ \int_{-W_A}^{W_1} \frac{\ln\left[1 + \exp\left(-\dfrac{W_x + \Phi}{k\,T}\right)\right]}{1 + \exp\left[\dfrac{8\pi\,(2\,m_e)^{1/2}\, v(y)\, W_x^{3/2}}{3\,h\,e\,E}\right]}\, dW_x \; + \right.$$

$$\left. + \int_{W_1}^{\infty} \ln\left[1 + \exp\left(-\frac{W_x + \Phi}{k\,T}\right)\right] dW_x \right\}$$

with

$$v(y) = \begin{cases} -\,(y/2)^{1/2}\left[-\,2\,\mathfrak{E}(k_1) + (1 + y)\,\mathfrak{K}(k_1)\right] & \text{for } y \geq 1 \\ (1 + y)^{1/2}\left[\mathfrak{E}(k_2) - y\,\mathfrak{K}(k_2)\right] & \text{for } y \leq 1 \end{cases}$$

$$\mathfrak{E}(k) = \int_0^{\pi/2} \left(1 - k^2 \sin^2\theta\right)^{1/2} d\theta$$

$$\mathfrak{K}(k) = \int_0^{\pi/2} \left(1 - k^2 \sin^2\theta\right)^{1/2} d\theta$$

$$k_1 = \left(\frac{y - 1}{2\,y}\right)^{1/2}$$

$$k_2 = \left(\frac{1 - y}{1 + y}\right)^{1/2}$$

$$y = \left(\frac{e^3\,E}{4\pi\,\varepsilon_0}\right)^{1/2} \frac{1}{W_x}$$

$$W_1 = -\left(\frac{e^3\,E}{8\pi\,\varepsilon_0}\right)^{1/2}$$

W_A is the height of the potential barrier with respect to the zero potential of electrons in the solid. Complex analytic solutions of the integrals are given by Murphy and Good [99]. For practical application, the formula

$$j^e_{TF} = j_F \left(\frac{\pi k T}{d_e}\right) / \sin\left(\frac{\pi k T}{d_e}\right)$$

with

$$d_e = \frac{h e E}{4\pi (2 m_e)^{1/2} \Phi^{1/2} t(y)}$$

is often used. It is a good approximation if $kT < \frac{2}{3} d_e$. The function t(y) was introduced above.

Additional simplified formulae which are, however, valid only in a limited T-E range can be found in [123], [124]. For example,

$$j^e_{TF}(T,E) = a \left(A T^2 + B E^{9/8}\right) \exp\left[-\left(T^2/C + E^2/D\right)^{-1/2}\right]$$

where a,A,B,C,D are constants.

References: [97] pp.23-28, [99], [100], [123], [124]

Remark:

Under certain circumstandes (e.g. in high-voltage vacuum systems) the field emission current, or thermofield emission current will lead to surface distortion of the emitting solid. The current will then increase by a factor 10^8 - 10^9 A/s (explosive emission).

References: [101], [125]

14.4. SECONDARY ELECTRON EMISSION (SEE)

SEE due to primary ion impact

The current density of secondary electrons is given by

$$j_{SEE} = \gamma_i \; j_i$$

where γ_i denotes the secondary electron yield per incident ion and j_i is the incidention ion current density. For singly charged, rare gas ions of negligible kinetic energy, the yield γ_i is tabulated below for some electrode materials [102]

ion	Ta	Mo	W	Ni
He^+	0.14	0.300	0.290	0.159
Ne		0.254	0.213	0.128
Ar	0.007	0.122	0.095	0.050
Kr		0.069	0.050	
Xe		0.022	0.013	0.005

For singly charged ions of other elements, values of γ_i are below 0.1.

SEE due to primary electron impact

The general energy dependence of the SEE yield is illustrated in the figure below.

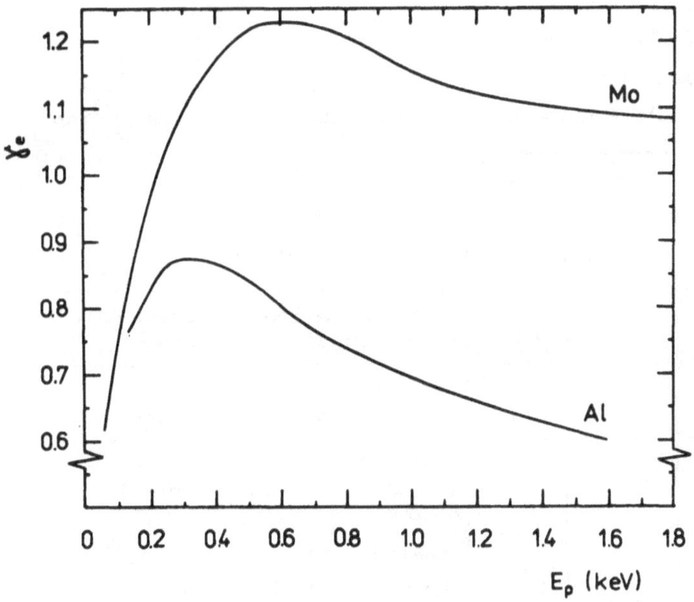

For some electrode materials, the maximum SEE yield, $\gamma_{e,max}$, caused by primary electrons of energy $E_{p,max}$ (in eV) is given in the following table:

Z	element	$\gamma_{e,max}$	$E_{p,max}$	Z	element	$\gamma_{e,max}$	$E_{p,max}$
6	C	1.0	200	47	Ag	1.70	800
13	Al	0.88	300	73	Ta	1.30	700
22	Ti	0.83	300	74	W	1.40	700
29	Cu	1.40	700	79	Au	1.81	900
42	Mo	1.23	600	80	Hg	1.63	700

References: [126] p.142, [175]

14.5. MODES OF ARC CATHODE OPERATION

All values given in the table below represent only the characteristic order of magnitude.

Nomenclature:

j : current density at the cathode surface

ϰ : characteristic erosion rate

$\delta U/U_c$: high-frequency voltage noise normalized by the cathode voltage drop

Non-stationary modes				
mode	j in A/m^2	ϰ in $\frac{\mu g}{As}$	$\delta U/U_c$	conditions, features
vapour mode type 1	10^{10} (average)	10	0.05	cold cathode, contaminated cathode surface, quickly moving spot
vapour mode type 2	10^{12}	100	0.5	cold cathode, clean cathode surface, quickly moving spot
high-current vapour mode type A	10^{12} in an individual spot	100		cold cathode, several quickly moving spots, inter-electrode plasma is diffuse
			0.3	passive anode
			1	footpoints at the anode
high-current vapour mode type B	10^8 (average)	1000-10000	0.5	cold cathode, single spots merge and form a large cathode spot, inter-electrode plasma column is contracted, anode spot

Stationary modes				
mode	j in A/m^2	ϰ in $\frac{\mu g}{As}$	$\delta U/U_c$	conditions, features
thermionic vapour mode	10^5 - 10^6	100-10000	$<10^{-3}$	strongly evaporating cathode, liquid cathode in vacuum, diffuse cathode spot
thermionic spot mode at low p	10^4 - 10^7	0.01	$<10^{-3}$	hot cathode with low work function in a gas at low pressure, diffuse cathode spot
thermionic spot mode at high p	10^7 - 10^8	0.001-0.1	$<10^{-3}$	hot cathode with low work function in a gas at high pressure, well defined cathode spot
thermionic spotless mode	10^7	0.00001	$<10^{-3}$	very high global cathode temperature, gas at high pressure, diffuse arc attachment

Reference: [127]

15. MISCELLANEOUS

15.1. CAPACITIES OF ELECTRODE CONFIGURATIONS

In this section

ε_r : relative permittivity (dielectric constant)

$$K_\varepsilon = \begin{cases} 1 / (4\pi \varepsilon_o) & (SI) \\ 1 & (CGS) \end{cases}$$

Other symbols can be identified with the aid of the figures to the right of each formula.

(i) two concentric spheres

$$C_{AB} = \frac{\varepsilon_r}{K_\varepsilon} \frac{r_B r_A}{r_B - r_A}$$

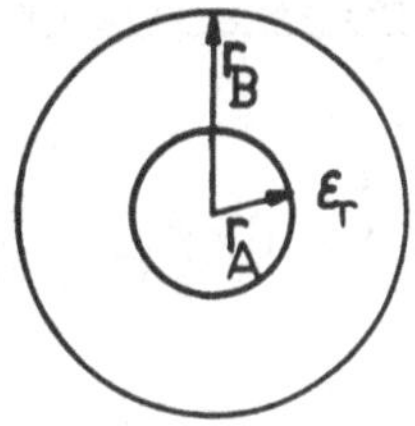

(ii) two small spheres

$$C_{AB} = \frac{\varepsilon_r}{4 K_\varepsilon} \frac{r (2a - r)}{a - r}$$

$$C_{AB} \xrightarrow[r \ll a]{} \frac{\varepsilon_r}{2 K_\varepsilon} r$$

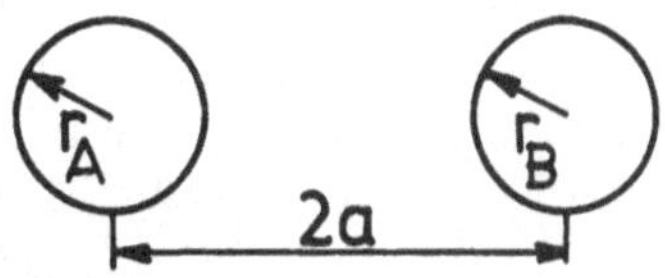

(iii) two long concentric cylinders $(l \gg r_B , r_A)$

$$C^*_{AB} = C_{AB}/l = \frac{\varepsilon_r}{4 K_\varepsilon \ln (r_B/r_A)}$$

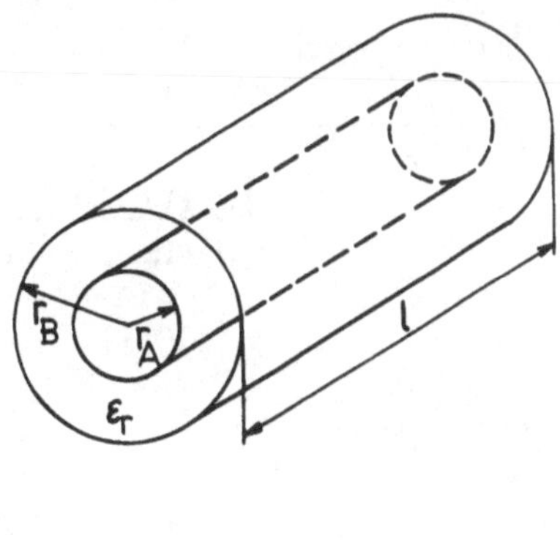

(iv) two long parallel cylinders

$$C^*_{AB} = C_{AB}/l = \frac{\varepsilon_r}{4\,K_\varepsilon}\left(\ln\left\{\frac{a}{r} + \left((a/r)^2 - 1\right)^{1/2}\right\}\right)^{-1}$$

$$C^*_{AB} \cong \frac{\varepsilon_r}{4\,K_\varepsilon}\left(\ln\left(\frac{2a}{r}\right)\right)^{-1}$$

for $l \gg a$ and $r \ll a$

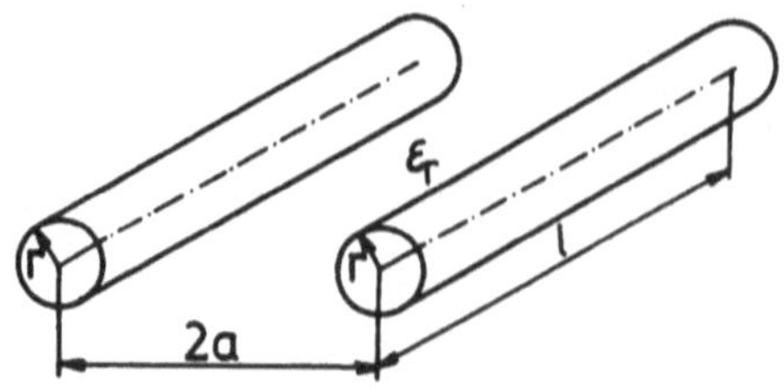

(v) very long cylinder - conducting plane

$$C^*_{AO} = C_{AO}/l = \frac{\varepsilon_r}{2\,K_\varepsilon}\left(\ln\left\{\frac{h}{r} + \left((h/r)^2 - 1\right)^{1/2}\right\}\right)^{-1}$$

$$C^*_{AB} \cong \frac{\varepsilon_r}{2\,K_\varepsilon}\left(\ln\left(\frac{2h}{r}\right)\right)^{-1}$$

for $l \gg h$ and $r \ll h$

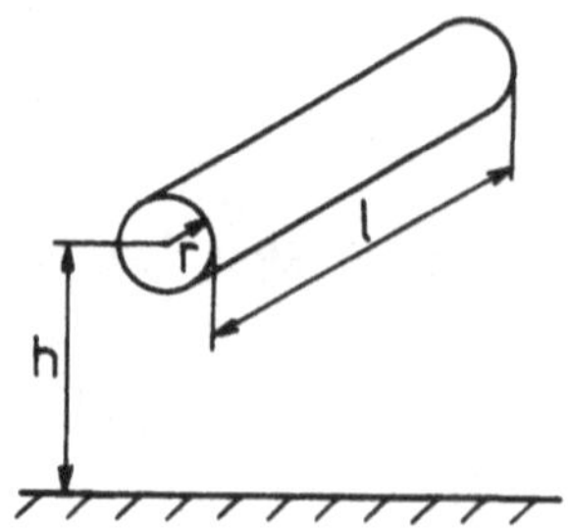

(vi) horizontal cylinder of finite length - conducting plane ($d \ll l$)

$$C_{AO} = \frac{\varepsilon_r}{2\,K_\varepsilon}\;\frac{1}{\ln\left[\frac{2\,l}{d}\left(\frac{(l^2+16\,h^2)^{1/2} - 1}{(l^2+16\,h^2)^{1/2} + 1}\right)^{1/2}\right]}$$

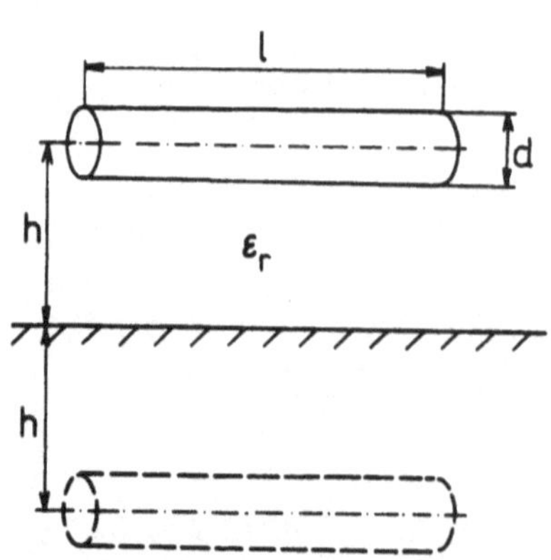

(vii) vertical cylinder of finite length - conducting plane (d « l)

$$C_{AO} = \frac{\varepsilon_r}{2\ K_\varepsilon} \frac{1}{\ln\left[\frac{2\ l}{d} \frac{(l + 4\ h)}{3\ l + 4\ h}\right]}$$

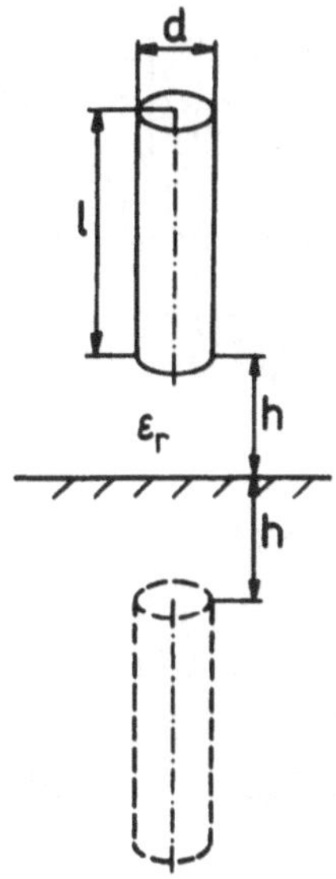

Reference: [104]

15.2. INDUCTANCES OF ELECTRODE CONFIGURATIONS

In this section

μ_r : relative permeability

$$K_\mu = \begin{cases} 4\pi/\mu_o & (SI) \\ 1 & (CGS) \end{cases}$$

(i) long conductor

$$L^* = L/l = \frac{\mu_{r,conduc}}{K_\mu}$$

(ii) two parallel conductors $(l \gg a \gg d_1\ ,\ d_2)$

$$L^* = L/l = \frac{1}{4\ K_\mu} \ln\left(\frac{4\ a}{\left(d_1\ d_2\right)^{1/2}}\right)$$

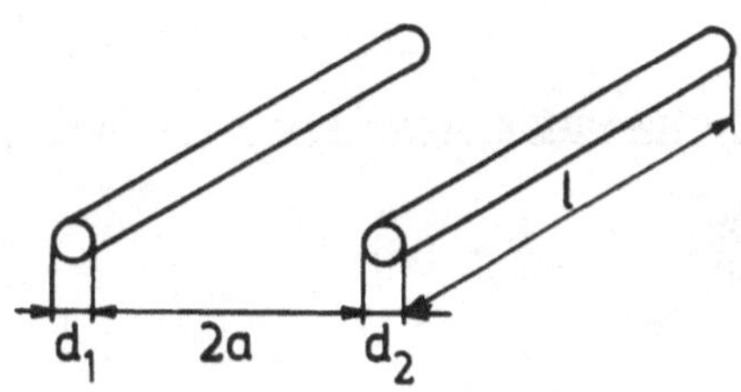

(iii) toroidal coil with circle cross section

$$L = \frac{4\pi\ \mu_r}{K_\mu} N^2 \left(R - \left(R^2 - r^2\right)^{1/2}\right)$$

where N is the number of turns

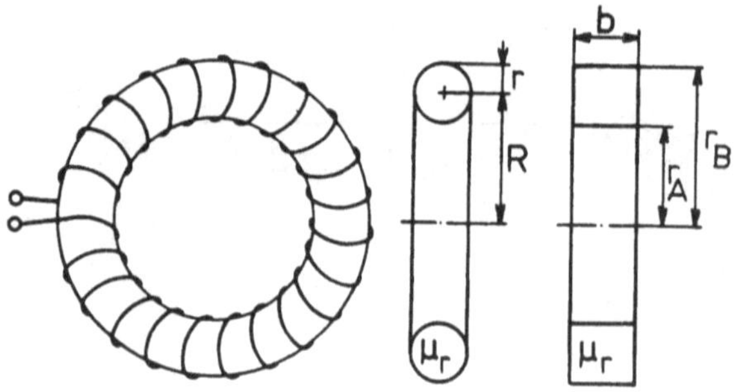

(iv) toroidal coil with rectangular cross section

$$L = \frac{\mu_r}{2\ K_\mu} N^2\ b\ \ln\left(r_B/r_A\right)$$

(v) conducting torus with circle cross section

$$L = \frac{4\pi}{K_\mu} R \left(\ln \frac{8\ R}{r} - 2\right)$$

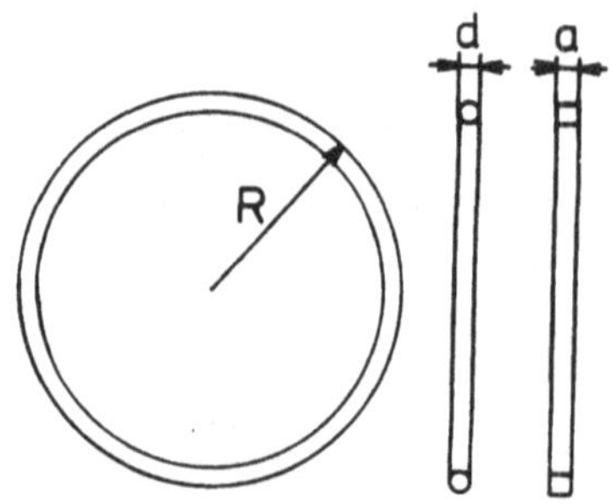

(vi) conducting torus with rectangular cross section (a < R)

$$L = \frac{4\pi}{K_\mu} R \left(\ln \frac{R}{a} + 1.5\right)$$

(vii) coaxial cable

$$L^* = \frac{L}{l} = \underbrace{\frac{\mu_{r1}}{2\,K_\mu}}_{1)} + \underbrace{\frac{2\,\mu_{r2}}{K_\mu}\,\ln\frac{r_B}{r_A}}_{2)} + \underbrace{\frac{2\,\mu_{r3}\,r_C^4}{K_\mu\,(r_C^2 - r_B^2)^2}\,\ln\left(\frac{r_C}{r_B} - \frac{(3\,r_C^2 - r_B^2)\,(r_C^2 - r_B^2)}{4\,r_C^4}\right)}_{3)}$$

1) inner conductor

2) insulator

3) outer conductor

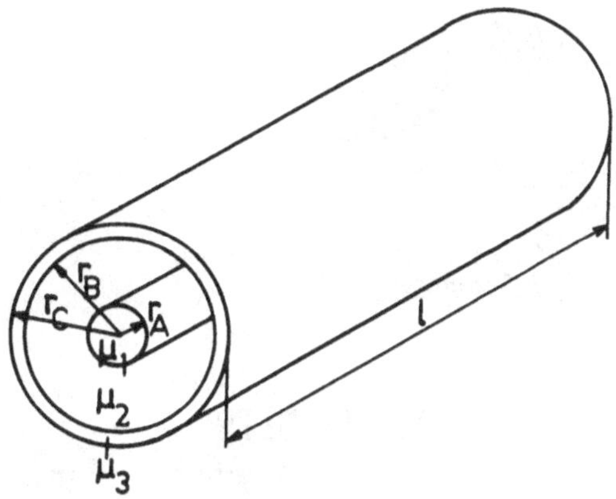

(viii) parallel strip lines

$$L^* = \frac{L}{l} = \frac{8\,\mu_r}{K_\mu}\,\ln\left(1 + \frac{a}{d + b}\right)$$

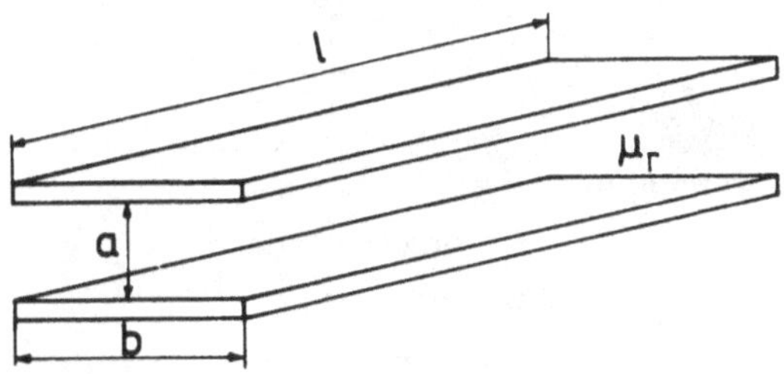

(ix) coil with circlular cross-section having a single layer of turns

N is the number of turns

$$L = \frac{\pi^2\,\mu_r}{K_\mu}\,N^2\,\frac{D^2}{\left(l^2 + D^2\right)^{1/2}}$$

for l = (5...10) D ;

$$L = \frac{\pi^2\,\mu_r}{K_\mu}\,N^2\,\frac{D}{l} \quad \text{for } l \gg D$$

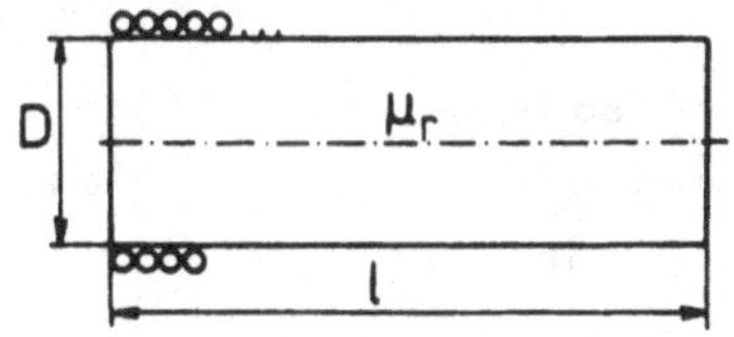

(x) coil with circlular cross-section having many layers of turns

$$L = \frac{\pi^2 \mu_r}{6 K_\mu} \frac{D_2^4 - 4 D_2 D_1^3 + 3 D_1^4}{\left(D_2 - D_1\right)^2} \qquad \text{for} \quad l \gg \frac{1}{2}\left(D_1 + D_2\right)$$

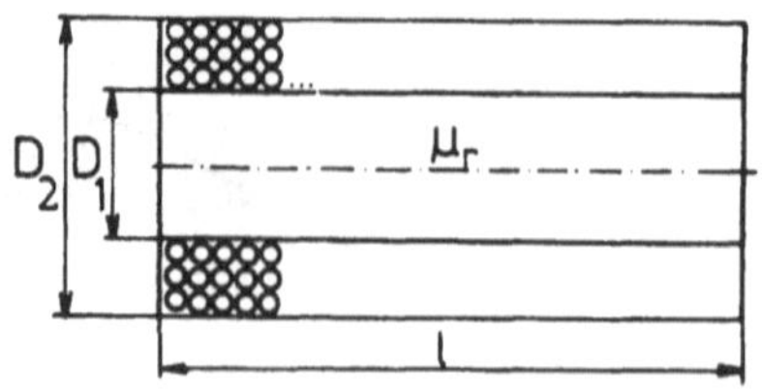

Reference: [104]

15.3. AC CIRCUITS

Circuit:

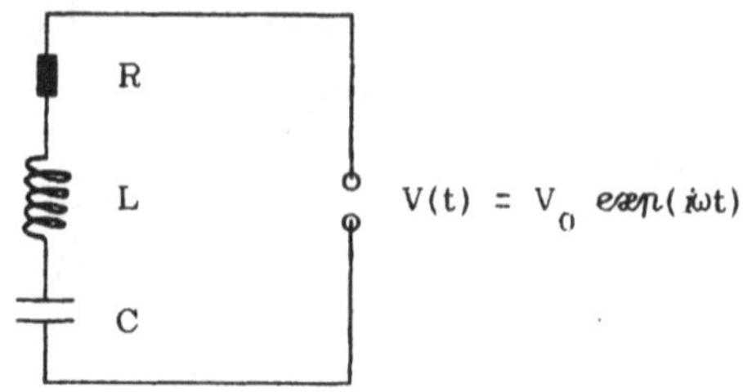

Notation:

R : ohmic resistance

L : inductance

C : capacitance

Q : charge

V : voltage

$I = dQ/dt$: current

Z : impedance, $i = (-1)^{1/2}$

Circuit equation:

$$\boxed{L \frac{d^2Q}{dt^2} + R \frac{dQ}{dt} + \frac{Q}{C} = V}$$

Solutions:

$$Q(t) = Q_{ss} + Q_t , \qquad I(t) = I_{ss} + I_t$$

with

$$I_{ss} = i\,\omega\, Q_{ss} = V/Z$$

$$I_t = \frac{dQ_t}{dt}$$

$$Z = R + i\,\omega\,L + \frac{1}{i\,\omega\,C} = R + i\left(\omega\,L - \frac{1}{\omega\,C}\right)$$

For initial conditions

$$Q(0) \equiv Q_0 = \overline{Q}_0 + Q_{ss}\ , \qquad I(0) \equiv I_0$$

the transients can be of three types, depending on $\Delta = R^2 - 4\,\frac{L}{C}$

<u>(i) overdamped, $\Delta > 0$</u>

$$Q_t = \frac{I_0 + \gamma_+\,\overline{Q}_o}{\gamma_+ - \gamma_-}\,exp\left(-\,\gamma_- t\right) - \frac{I_0 + \gamma_-\,\overline{Q}_o}{\gamma_+ - \gamma_-}\,exp\left(-\,\gamma_+ t\right)$$

$$I_t = \frac{\gamma_+\,(I_0 + \gamma_-\,\overline{Q}_o)}{\gamma_+ - \gamma_-}\,exp\left(-\,\gamma_+ t\right) - \frac{\gamma_-\,(I_0 + \gamma_+\,\overline{Q}_o)}{\gamma_+ - \gamma_-}\,exp\left(-\,\gamma_- t\right)$$

where

$$\gamma_\pm = \frac{R \pm \Delta^{1/2}}{2\,L}$$

<u>(ii) critically damped, $\Delta = 0$</u>

$$Q_t = \left[\overline{Q}_0 + \left(I_0 + \gamma_r\,\overline{Q}_0\right)\,t\right]\,exp\left(-\,\gamma_r t\right)$$

$$I_t = \left[I_0 - \left(I_0 + \gamma_r\,\overline{Q}_0\right)\,\gamma_r t\right]\,exp\left(-\,\gamma_r t\right)$$

where $\gamma_r = R\,/2L$

<u>(iii) underdamped, $\Delta < 0$</u>

$$Q_t = \left[\frac{\gamma_r\,\overline{Q}_0 + I_0}{\omega_1}\,sin(\omega_1 t) + \overline{Q}_0\,cos(\omega_1 t)\right]\,exp\left(-\,\gamma_r t\right)$$

$$I_t = \left[I_0\,cos(\omega_1 t) - \frac{\left(\omega_1^2 + \gamma_r^2\right)\,\overline{Q}_0 + \gamma_r\,I_0}{\omega_1}\,sin(\omega_1 t)\right]\,exp\left(-\,\gamma_r t\right)$$

where

$$\omega_1 = \omega_0\left(1 - \frac{R^2\,C}{4\,L}\right)^{1/2} \qquad \text{and} \qquad \omega_0 = \left(L\,C\right)^{1/2}$$

When $\omega = \omega_0$ (i.e. reseonance) then $Z = R$.

The quality of the circuit is defined by $\Omega = \omega_0 L/R$.

<u>Reference</u>: [35]

15.4. WAVELENGTHS OF LASERS

type of laser	wavelength in nm	cw	pulsed
CO_2	10,600	yes	yes
CO	5,000	yes	yes
Holmium	2,060	yes	yes
Iodine	1,315	no	yes
Nd-YAG	1,060	yes	yes
Ruby	694.3	yes	yes
Kr ion	676.1	yes	yes
	647.1	yes	yes
He-Ne	632.8	yes	no
Au vapour	628	no	yes
Cu vapour	510	no	yes
Ar + Kr ion	647.1	yes	yes
	514.5	yes	yes
	488.0	yes	yes
Ar ion	514.5	yes	yes
	488.0	yes	yes
	476.2	yes	yes
	457.9	yes	yes
XeF	353	no	yes
N_2	337.1	no	yes
XeCl	308	no	yes
KrF	248	no	yes
KrCl	222	no	yes
ArF	193	no	yes
Xe	175	no	yes
dye (tunable)	1280 - 320	yes	yes

References: There are a vast amount of literature on lasers, we give only a few references which could helpful for introducing in this matter: [183]-[188].

15.5. DENSITY-TEMPERATURE PLANE FOR VARIOUS PLASMAS

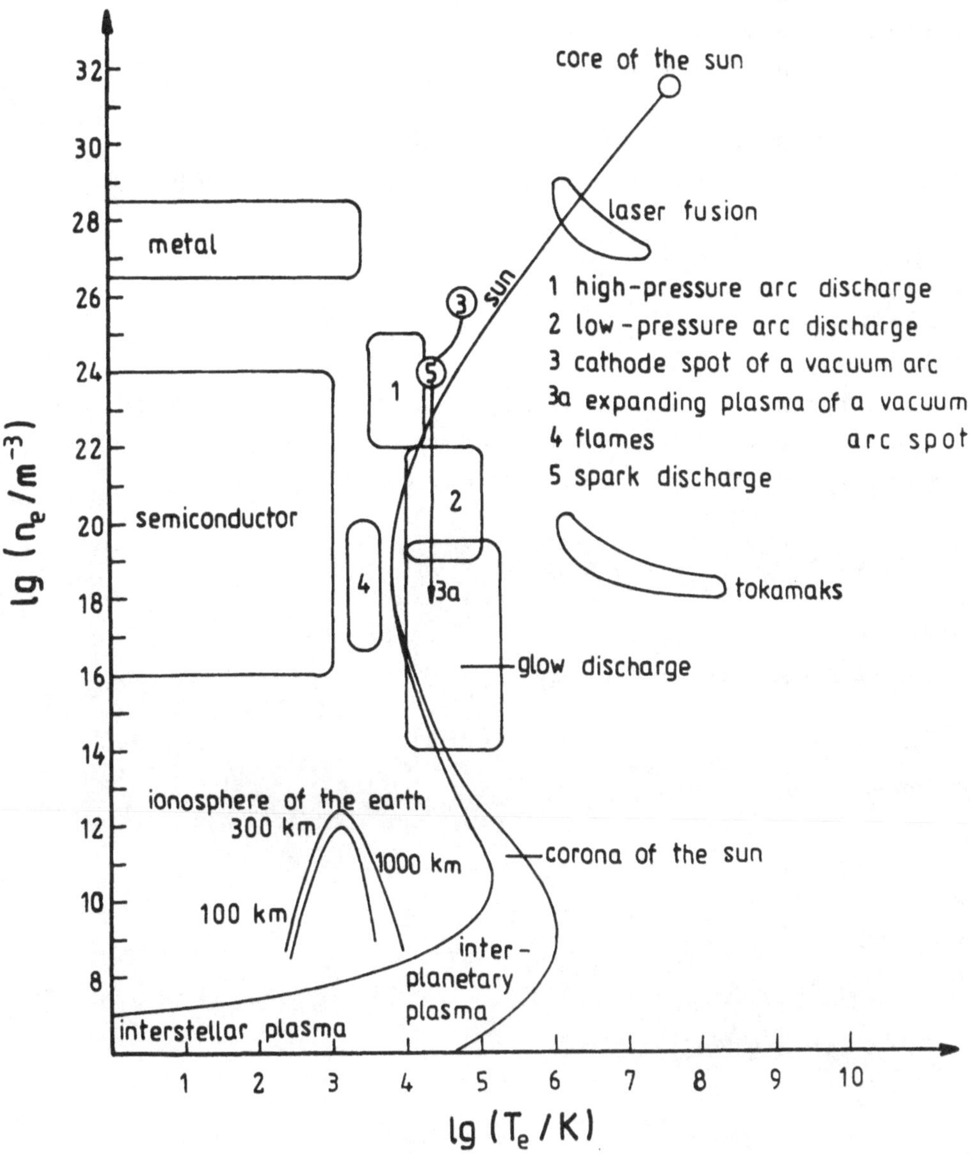

16. MATHEMATICS

16.1. VECTOR CALCULUS

Notation: f,g,...scalars

A,B,...vectors

grad $f \equiv \nabla f$, div $A \equiv \nabla \circ A$, rot $A \equiv$ curl $A \equiv \nabla \times A$

$$A \circ (B \times C) = B \circ (C \times A) = C \circ (A \times B) \equiv (\, A\ B\ C\,)$$

$$A \times (B \times C) = B\,(A \circ C) - C\,(A \circ B)$$

$$A \times (B \times C) + B \times (C \times A) + C\,(A \times B) = 0$$

$$(A \times B) \circ (C \times D) = (A \circ C)\,(B \circ D) - (A \circ D)\,(B \circ C)$$

$$(A \times B) \times (C \times D) = (A\ B\ D)\,C - (A\ B\ C\,)\,D = (A\ C\ D)\,B - (B\ C\ D)\,A$$

$$\nabla\,(fg) = f\,\nabla g + g\,\nabla f$$

$$\nabla \circ (fA) = f\,\nabla \circ A + A \circ \nabla f$$

$$\nabla \times (fA) = f\,\nabla \times A + \nabla f \times A$$

$$\nabla \circ (A \times B) = B \circ (\nabla \times A) - A \circ (\nabla \times B)$$

$$\nabla \times (A \times B) = A\,(\nabla \circ B) - B\,(\nabla \circ A) + (B \circ \nabla)\,A - (A \circ \nabla)\,B$$

$$\nabla \circ (A \circ B) = A \times (\nabla \times B) + B \times (\nabla \times A) + (A \circ \nabla)\,B + (B \circ \nabla)\,A$$

$$(A \circ \nabla)\,A = \nabla \left(\frac{A^2}{2} \right) - A \times (\nabla \times A)$$

$$\nabla^2 f \equiv \Delta f = \nabla\,(\nabla f)$$

$$\nabla^2 A \equiv \Delta A = \nabla\,(\nabla \circ A) - \nabla \times (\nabla \times A)$$

$$\nabla \times \nabla f = 0$$

$$\nabla \circ (\nabla \times A) = 0$$

$$r = xi + yj + zk$$

$$\nabla \circ r = 3$$

$$\nabla \times r = 0$$

$$(A \circ \nabla)\,r = A$$

If V is a volume enclosed by the surface S and $d\mathbf{S} = \mathbf{n}\, dS$, where n is the unit normal outward from V,

$$\int_{\text{closed volume}} dV\, \nabla f = \int_{\text{bounding surface}} dS\, \mathbf{n}\, f$$

$$\int_{\text{closed volume}} dV\, \nabla \circ \mathbf{A} = \int_{\text{bounding surface}} dS\, \mathbf{n} \circ \mathbf{A} \qquad \text{Gauss's theorem}$$

$$\int_{\text{closed volume}} dV\, \nabla \times \mathbf{A} = \int_{\text{bounding surface}} dS\, \mathbf{n} \times \mathbf{A}$$

If S is an open surface bounded by the contour C (dl is a line element),

$$\int_{\text{open surface}} dS\, \mathbf{n} \circ \nabla \times \mathbf{A} = \oint_{\text{bounding circumference}} d\mathbf{l} \circ \mathbf{A} \qquad \text{Stokes's theorem}$$

$$\int_{\text{open surface}} dS\, \mathbf{n} \times \nabla f = \oint_{\text{bounding circumference}} d\mathbf{l}\, f$$

Cylindrical coordinates (r,θ,z)

gradient $$\nabla f = \frac{\partial f}{\partial r}\,\vec{r} + \frac{1}{r}\frac{\partial f}{\partial \theta}\,\vec{\theta} + \frac{\partial f}{\partial z}\,\vec{z}$$

divergence $$\nabla \circ A = \frac{1}{r}\frac{\partial}{\partial r}\left(r\, A_r\right) + \frac{1}{r}\frac{\partial}{\partial \theta}\left(A_\theta\right) + \frac{\partial}{\partial z}\left(A_z\right)$$

curl $$\nabla \times A = \left[\frac{1}{r}\frac{\partial A_z}{\partial \theta} - \frac{\partial A_\theta}{\partial z}\right]\vec{r} + \left[\frac{\partial A_r}{\partial z} - \frac{\partial A_z}{\partial r}\right]\vec{\theta} + \\ + \left[\frac{1}{r}\frac{\partial}{\partial r}\left(r\, A_\theta\right) - \frac{1}{r}\frac{\partial}{\partial \theta}\left(A_r\right)\right]\vec{z}$$

Laplacian $$\nabla^2 f = \frac{1}{r}\frac{\partial}{\partial r}\left(r\,\frac{\partial f}{\partial r}\right) + \frac{1}{r^2}\frac{\partial^2 f}{\partial \theta^2} + \frac{\partial^2 f}{\partial z^2}$$

$$\nabla^2 A = (\nabla \circ \nabla)\, A = \left[\nabla^2 A_r - \frac{1}{r^2}\left(A_r + 2\,\frac{\partial A_\theta}{\partial \theta}\right)\right]\vec{r} + \\ + \left[\nabla^2 A_\theta - \frac{1}{r^2}\left(A_\theta - 2\,\frac{\partial A_r}{\partial \theta}\right)\right]\vec{\theta} + \nabla^2 A_z\,\vec{z}$$

$$(A \circ \nabla)\, B = \left[A_r \frac{\partial B_r}{\partial r} + A_\theta \frac{1}{r}\frac{\partial B_r}{\partial \theta} + A_z \frac{\partial B_r}{\partial z} - \frac{1}{r} A_\theta B_\theta\right] \vec{r} +$$

$$+ \left[A_r \frac{\partial B_\theta}{\partial r} + A_\theta \frac{1}{r}\frac{\partial B_\theta}{\partial \theta} + A_z \frac{\partial B_\theta}{\partial z} + \frac{1}{r} A_\theta B_r\right] \vec{\theta} +$$

$$+ \left[A_r \frac{\partial B_z}{\partial z} + A_\theta \frac{1}{r}\frac{\partial B_z}{\partial \theta} + A_z \frac{\partial B_z}{\partial z}\right] \vec{z}$$

<u>Spherical</u> <u>coordinates</u> (r,θ,ϕ)

gradient $$\nabla f = \frac{\partial f}{\partial r}\vec{r} + \frac{1}{r}\frac{\partial f}{\partial \theta}\vec{\theta} + \frac{1}{r \sin\theta}\frac{\partial f}{\partial \phi}\vec{\phi}$$

divergence $$\nabla A = \frac{1}{r^2}\frac{\partial}{\partial r}\left(r^2 A_r\right) + \frac{1}{r\sin\theta}\frac{\partial}{\partial\theta}\left(A_\theta \sin\theta\right) + \frac{1}{r\sin\theta}\frac{\partial A_\phi}{\partial\phi}$$

curl $$\nabla \times A = \left[\frac{1}{r\sin\theta}\frac{\partial}{\partial\theta}\left(A_\phi \sin\theta\right) - \frac{1}{r\sin\theta}\frac{\partial A_\theta}{\partial\phi}\right]\vec{r} +$$

$$+ \left[\frac{1}{r\sin\theta}\frac{\partial A_r}{\partial\phi} - \frac{1}{r}\frac{\partial}{\partial r}\left(r A_r\right)\right]\vec{\theta} +$$

$$+ \left[\frac{1}{r}\frac{\partial}{\partial r}\left(r A_\theta\right) - \frac{1}{r}\frac{\partial A_r}{\partial\theta}\right]\vec{\phi}$$

Laplacian $$\nabla^2 f = \frac{1}{r^2}\frac{\partial}{\partial r}\left(r^2\frac{\partial f}{\partial r}\right) + \frac{1}{r^2\sin\theta}\frac{\partial}{\partial\theta}\left(\sin\theta\frac{\partial f}{\partial\theta}\right) + \frac{1}{r^2\sin^2\theta}\frac{\partial^2 f}{\partial^2\phi}$$

$$\nabla^2 A = \left[\nabla^2 A_r - \frac{2A_r}{r^2} - \frac{2}{r^2}\frac{\partial A_\theta}{\partial\theta} - \frac{2A_\theta\cot\theta}{r^2} - \frac{2}{r^2\sin\theta}\frac{\partial A_\phi}{\partial\phi}\right]\vec{r} +$$

$$+ \left[\nabla^2 A_\theta + \frac{2}{r^2}\frac{\partial A_r}{\partial\theta} - \frac{A_\theta}{r^2\sin^2\theta} - \frac{2\cos\theta}{r^2\sin^2\theta}\frac{\partial A_\phi}{\partial\phi}\right]\vec{\theta} +$$

$$+ \left[\nabla^2 A_\phi - \frac{A_\phi}{r^2\sin^2\theta} + \frac{2}{r^2\sin\theta}\frac{\partial A_r}{\partial\phi} + \frac{2\cos\theta}{r^2\sin^2\theta}\frac{\partial A_\phi}{\partial\phi}\right]\vec{\phi}$$

<u>References</u>: [35], [60], [128] appendix C, [129] appendix 1, [131] p.288

16.2. PLASMA DISPERSION FUNCTION

The plasma dispersion function is defined by

$$Z(\zeta) = \pi^{-1/2} \int_{-\infty}^{\infty} \frac{\exp\left(-t^2\right)}{t-\zeta}\, dt = 2\, i \exp\left(-\zeta^2\right) \int_{-\infty}^{i\zeta} \exp\left(-t^2\right) dt$$

where the first form is valid only for $\mathcal{Im}\ \zeta > 0$. Physically, $\zeta = x + i\,y$ is the ratio of wave phase velocity to thermal velocity.

The plasma dispersion function satisfies the differential equation

$$\frac{dZ}{d\zeta} = -2\,(1 + \zeta\, Z)\,, \qquad Z(0) = i\,\pi^{1/2}$$

$$\frac{d^2Z}{d\zeta^2} + 2\,\zeta\,\frac{dZ}{d\zeta} + 2\,Z = 0$$

The real argument can be written as (y=0)

$$Z(x) = \exp\left(-x^2\right)\left(i\,\pi^{1/2} - 2\int_0^x \exp\left(-t^2\right) dt\right)$$

and the imaginary argument (x=0)

$$Z(iy) = i\,\pi^{1/2} \exp\left(y^2\right)\left[1 - \mathrm{erf}(y)\right]$$

For small arguments, expansion into a power series gives

$$Z(\zeta) = i\,\pi^{1/2} \exp\left(-\zeta^2\right) - 2\,\zeta\left(1 - 2\,\zeta^2/3 + 4\,\zeta^4/15 - 8\,\zeta^6/105 + \ldots\right)$$

For large arguments, $|\zeta| \gg 1$, the function can be approximated by the asymptotic series [181]

$$Z(\zeta) = i\,\pi^{1/2}\,\sigma \exp\left(-\zeta^2\right) - \zeta^{-1}\left(1 + 1/2\zeta^2 + 3/4\zeta^4 + 15/8\zeta^6 + \ldots\right)$$

where

$$\sigma = \begin{cases} 0 & \text{for } y > |x|^{-1} \\ 1 & \text{for } |y| < |x|^{-1} \\ 2 & \text{for } y < -|x|^{-1} \end{cases}$$

Useful relations are

$$Z(\zeta^*) = -\left[Z(-\zeta)\right]^*$$

and

$$Z(\zeta^*) = \left[Z(\zeta)\right]^* + 2\, i\, \pi^{1/2}\, exp\left[-\left(\zeta^*\right)^2\right], \quad y > 0$$

The two-pole approximations [182]

$$Z(\zeta) \cong \frac{0.50 + 0.81\, i}{a - \zeta} - \frac{0.50 - 0.81\, i}{a^* - \zeta}, \quad a = 0.51 - 0.81\, i$$

$$Z'(\zeta) \cong \frac{0.50 + 0.96\, i}{(b - \zeta)^2} - \frac{0.50 - 0.96\, i}{(b^* - \zeta)^2}, \quad b = 0.48 - 0.91\, i$$

are good for ζ in upper half plane except when $y < \pi^{1/2}\, x^2\, exp\left(-\, x^2\right)$ for $x \gg 1$.

References: [35] p.30, [180]-[182]

REFERENCES

[1] E.R.Cohen, B.N.Taylor, *The 1986 CODATA recommended values of the fundamental physical constants*, J. Phys. Chem. Data 17 no.4 (1988) 1795 - 1803;
E.R.Cohen, B.N.Taylor, *The fundamental physical constants*, Physics Today, August 1988, BG9-BG13;
E.R.Cohen, B.N.Taylor, *The adjustment of the fundamental physical constants*, Report of the CODATA Task Group on Fundamental Constants, CODATA Bull. 63, Pergamon, Elmsford, New York, 1986

[2] F.Cabannes, J.Chapelle, *Spectroscopic plasma diagnostics*, chapter 7 of: *Reactions under plasma conditions*, vol.I, ed. by M.Venugopalan, Wiley-Interscience, New York, London, Sydney, Toronto, 1971

[3] M.L.Biberman, G.E.Norman, *On the calculation of photoionization absoption*, Optika i Spektrosk. 8 no.4 (1960) 433-438 (Opt. Spectrosc. 8 (1960) 230;
M.L.Biberman, G.E.Norman, K.N.Ul'yanov, *The photoionization of excited complex atoms and ions*, Astron. Zh. (USSR) 39 no.1 (1962) 105-115 (Sov. Astron.-AJ (USA) 6 no.1 (1962) 77-83; see also: Opt. Spectrosc. 10 (1961) 267

[4] D.Schlüter, *The calculation of bound-free transition probabilities and their application to continuum spectra of the rare gases*, J. Quant. Spectrosc. Radiative Transfer 5 (1965) 87-90; see also: Z. Astrophys. 61 (1965) 57

[5] T.Kihara, O.Aono, Y.Itikawa, *Unified theory of relaxations in plasmas. II. Applications*, J. Phys. Soc. Jap. 18 no.7 (1963) 1043-1050

[6] T.Kihara, O.Aono, *A review of the unified theory of relaxations in plasmas*, in: *Kinetic equations*, ed. by R.L.Liboff, N.Rostoker, 1971

[7] R.S.Cohen, P.McR.Routly, L.Spitzer, *The electrical conductivity of an ionized gas*, Phys. Rev. 80 no.2 (1950) 230-238

[8] B.M.Smirnov, *Physics of slightly ionized gases. Excercises and solutions* (in Russ.), Nauka, Moscow, 1985

[9] H.W.Drawin, *Influence of atom-atom collisions on the collisional-radiative ionization and recombination coefficients of hydrogen plasmas*, Z. Physik 225 (1969) 483-493

[10] J.J Thomson, Phil. Mag. 23 (1912) 449-457

[11] G.Elwert, *Über die Ionisierungs- und Rekombinationsprozesse in einem Plasma und die Ionisationsformel der Sonnenkorona*, Z. Naturforschung 7a (1952) 432-439

[12] M.Gryzinski, *Classical theory of electronic and ionic inelastic collisions*, Phys. Rev. 115 no.2 (1959) 374-383

[13] R.C.Stabler, *Classical impulse approximation for inelastic electron-atom collisions*, Phys. Rev. 123 no.5A (1964) A1268-A1273

[14] R.F.Post, *Impurity radiation losses from a high temperature plasma*, Plasma Physics 3 (1961) 273-286

[15] H.W.Drawin, *Zur formelmäßigen Darstellung der Ionisierungsquerschnitte gegenüber Elektronenstoß*, Z.Physik 164 (1961) 513-521

[16] M.J.Seaten, Planet. Space Science 12 (1964) 55-73

[17] M.Gryzinski, *Two-particle collisions*, Phys. Rev. 138 no.2A (1965) A305-A358

[18] R.H.McFarland, *Electron-impact ionization measurments of surface-ionizable atoms*, Phys. Rev. 159 no.1 (1967) 20-26

[19] L.Vriens, *Binary-encounter electron-atom collision theory*, Phys. Rev. 141 no.1 (1966) 88-92; Proc. Phys. Soc. 89 (1966) 13-21

[20] A.Burgess, Proc. 3rd Int. Conf. Phys. Elec. Atom. Coll., London 1963, 237-242; A.Burgess, M.J.Seaton, *A general formula for the calculation of atomic photo-ionization cross-sections*, Mont. Not. Roy. Astron. Soc. 120 no. 2 (1960) 121-151

[21] W.Lotz, *An empirical formula for the electron-impact ionization cross-section*, Z. Physik 206 (1967) 205-211; Astrophys. J. Suppl. 14 (1967) 207-238

[22] W.Lotz, Z. Physik 216 (1968) 241-247

[23] W.Lotz, *Electron-impact ionization cross-section for atoms up to Z = 108*, Z. Physik 232 (1970) 101-107

[24] L.A.Vainshtein, I.I.Sobelman, E.A.Yukov, *Electron-excitation cross-sections of atoms and ions* (in Russ.), Nauka, Moscow 1973

[25] K.L.Bell, H.B.Gilbody, J.G.Hughes, A.E.Kingston, F.J.Smith, J. Phys. Chem. Ref. Data 12 (1983) 891-916; see also: Culham Report CLM-R 216 (1982)

[26] T.D.Märk, G.H.Dunn (Ed.), *Electron impact ionization*, Springer-Verlag Wien, New York 1985

[27] H.W.Drawin, P.Felenbok, *Data for plasmas in local thermodynamic equilibrium*, Gauthier-Villars, Paris 1965

[28] A.Unsöld, *Physik der Sternatmosphären*, 2nd ed., Springer-Verlag Berlin, Heidelberg, New York 1956; Corrected Reprint of the 2nd ed. Springer-Verlag Berlin, Heidelberg, New York 1968

[29] Ch.E.Moore, *Atomic energy levels*, vol. I, II, III, Circular of the National Bureau of Standards No. 467, Washington DC, 1949, 1952, 1958

[30] L.A.Dushin, O.S.Pavlichenko, *Plasma diagnostics with lasers* (in Russ.), Atomizdat, Moscow 1968, German transl.: Akademie-Verlag, Berlin, 1973

[31] L.Spitzer, *Physics of fully ionzed gases*, Interscience Publishers, New York, 1956; Russ. transl.: IL Moscow, 1957, and Mir, Moscow, 1965

[32] M.Mitchner, Ch.H.Kruger Jr., *Partially ionized gases*, John Wiley & Sons, New York, London, Sydney, Toronto 1973

[33] L.M.Biberman, V.S.Vorobev, I.T.Yakubov, *Kinetics of the non-equilibrium low temperature plasma* (in Russ.), Nauka, Moscow 1982

[34] W.L.Wiese, M.W.Smith, B.M.Glennon, *Atomic transition probabilities*, NBS Circular, National Bureau of Standards, Washington DC, vol.I: Hydrogen trough neon (1966), vol.: Sodium through calcium (1969)

[35] D.L.Book, *NRL Plasma formulary*, 1987 revised edition, Naval Research Laboratory, NRL Publication 0084-4040, Washington DC, 1987

[36] V.E.Golant, *Microwave methods in plasma diagnostics* (in Russ.), Nauka Moscow 1968

[37] F.Cap, *Einführung in die Plasmaphysik*.
vol.1: *Theoretische Grundlagen*
vol.2: *Wellen und Instabilitäten*
vol.3: *Magnetohydrodynamik*, Akademie-Verlag, Berlin and Pergamon Press, Oxford, Vieweg & Sohn, Braunschweig 1972

[38] S.-I Pai, *Magnetogasdynamics and plasma dynamics*, Springer-Verlag Wien, 1962

[39] J.C.M.DeHaas, *Non-equilibrium in flowing atmospheric plasmas*, Ph.D.Thesis, Tech. Univ. Eindhoven 1986

[40] V.E.Fortov, I.T.Yakubov, *Physics of non-ideal plasmas* (in Russ.), IVTAN, Chernogolovka 1984

[41] K.A.Günther, R.Radtke, *Electrical properties of weakly nonideal plasmas*, Akademie-Verlag Berlin, Birkhäuser Verlag Basel, Boston, Stuttgart 1984

[42] M.A.Heald, C.B.Wharton, *Plasma diagnostics with microwaves*, John Wiley & Sons Inc. New York, London, Sydney 1965;
C.B.Wharton, *Plasma diagnostics with microwaves*, chpt.11 of: *Plasma diagnostic techniques*, ed. by R.H.Huddlestone and S.L.Leonard, Academic Press, New York, 1965; Russian Transl.: Mir, Moscow, 1967

[43] L.Oster, *Emission, absorption, and conductivity of a fully ionized gas at radio frequencies*, Rev. Mod. Phys. 33 no.4 (1961) 525-543

[44] J.Dawson, C.Oberman, *High-frequency conductivity and the emission and absorption coefficients of a fully ionized plasma*, Phys. Fluids 5 no.5 (1962) 517-524

[45] W.Ebeling, W.D.Kremp, D.Kraeft, *Theory of bound states and ionization equilibrium in plasmas and solids*, Akademie-Verlag, Berlin 1976, Russ. transl.: Mir, Moscow 1979

[46] S.-I Pai, *Radiation gasdynamics*, Springer-Verlag, Wien, New York, 1966

[47] S.Chandrasekhar, *Radiative transfer*, Clarendon Press, Oxford, 1950

[48] V.V.Ivanov, *Transfer of radiation in spectral lines*, NBS Special Publication 385 (Nov. 1973),(English language edition of: *Radiative Transfer and the spectra of celestial bodies* (in Russian), Moscow, 1969)

[49] H.A.Kramers, *On the theory of X-ray absorption and the continuous X-ray spectrum*, Phil. Mag. 46 no.275 (1923) 836

[50] A.Unsöld, Ann. Phys. 33 (1938) 607

[51] H.Maecker, T.Peters, *Das Elektronenkontinuum in der Säule des Hochstrom kohlebogens und in anderen Bögen*, Z. Physik 139 no.4 (1954) 448-463

[52] W.Neumann, *Spektroskopische Methoden der Plasmadiagnostik*, in: *Ergebnisse der Plasmaphysik und Gaselektronik*, Band 1, ed. by R.Rompe, M.Steenbeck, Akademie-Verlag Berlin 1967

[53] H.Schirmer, *Zur Theorie der Plasmen unter besonderer Berücksichtigung der Xenon-Hochdruckentladung*, Habilitationsschrift, Tech. Univ. Berlin, 1960

[54] A.Unsöld, B.Baschek, *Der neue Kosmos*, Springer-Verlag, Berlin Heidelberg, New York, 3rd ed. 1981, 4th ed. 1988

[55] A.C.G.Mitchell, M.W.Zemansky, *Resonance radiation and excited atoms*, University Press, Cambride 1961 (first ed. 1934)

[56] H.R.Griem, *Plasma Spectroscopy*, McGraw-Hill Book Company, New York 1964, Russ. transl.: Atomizdat, Moscow 1969, 1975

[57] H.R.Griem, *Spectral line broadening by plasmas*, Academic Press, New York, London, Russ. transl.: Mir, Moscow 1978

[58] I.F.Golodnev et al., *Radiation heat transfer in high temperature gases* (in Russ.), ed. by R.I.Soloukhin, Energoatomizdat, Moscow 1984

[59] L.N.Pyatnitskii, *Plasma diagnostics with lasers* (in Russ.), Atomizdat, Moscow, 1976

[60] F.F.Chen, *Introduction to Plasma Physics abd controlled fusion*, 2nd ed., vol.1: *Plasma physics*, Plenum Press New York, London 1984, Russ. transl.: Mir, Moscow 1987

[61] D.A.Frank-Kamenetskii, *Lectures in plasma physics* (in Russ.), Atomizdat, Moscow 1968

[62] L.A.Artsimovich, *Elementary plasma physics*, Atomizdat, Moscow 1969

[63] H.W.Drawin, *Thermodynamic properties of the equilibrium and nonequilibrium states of plasmas*, in: *Reactions under plasma conditions*, ed by M.Venugopalan vol.1, Wiley Interscience New York, London, Sydney, Toronto 1971

[64] G.Joos, *Lehrbuch der theoretischen Physik*, Akademische Verlagsgesellschaft Geest & Portig, Leipzig 1954

[65] C.Królikowski (Ed.), *Podstawy wyładowan i wytrzymałosc elektryczna w próżni* (Fundamentals of discharges and electrical insulation in vacuum), Panstwoe Wydawnictwo Naukowe,Warszawa, Poznan 1983

[66] S.Chapman, T.Cowling, *The mathematical theory of non-uniform gases*, Cambride University Press 1964

[67] G.Sutton, A.Sherman, *Engineering magnetohydrodynamics*, McGraw-Hill, New York, 1965

[68] A.Rutscher, H.Deutsch (Ed.), *Wissensspeicher Plasmatechnik*, Fachbuchverlag, Leipzig, 1983

[69] W.Thomson, *An introduction to plasma physics*, Pergamon Press, 1962

[70] G.Schmidt, *Physics of high temperature plasma*, Academic Press, New York, 1966

[71] N.van Kampen, B.Felderhof, *Theoretical methods in plasma physics*, North Holland, 1967

[72] R.Balescu, *Irreversible processes in ionized gases*, Phys. Fluids 3 no.1 (1960) 52-63

[73] R.Balescu, *Statistical mechanics of charged particles*, Interscience Publishers, New York 1973

[74] L.Spitzer, R.Härm, *Transport phenomena in a completely ionized gas*, Phys. Rev. 89 no.5 (1953) 977-981

[75] V.L.Ginzburg, A.A.Rukhadze, *Waves in a magnetoactive plasma* (in Russian), Nauka, Moscow 1970

[76] U.Ascoli-Bartoli, A.deAngelis, S.Martellucci, *Wavelength dependence of the refractive index of a plasma in the optical region*, Nuovo Cimento 18 no.6 (1960) 1116-1137

[77] D.R.White, *Optical refractivity of high-temperature gases*, Phys. Fluids 4 no.1 (1961) 40-45

[78] C.W.Allen, *Astrophysical quantities*, The Atholone Press, London, 2nd ed. 1963

[79] I.Ursu, D.Apostol, M.Stoica, I.Apostol, D.Cracium, A.Hennig, I.N.Mihailescu, V.Stancalie, *Laser interferometry at* 10.6 μm *for plasma diagnostics*, Proc. 2nd Conf. on Trends in Quantum Electronics, ed. by A.M.Prokhorov, I.Ursu, Springer-Verlag, Berlin, Heidelberg, New York, London, Paris, Tokyo 1986, 443-462

[80] W.C.Marlow, *Hakenmethode* (in Engl.), Applied Optics 6 no.10 (1967) 1715-1724

[81] V.G.Dulov, G.A.Luk'yanov, *Gasdynamic processes of exhaustion* (in Russian), Nauka, Novosibirsk, 1984

[82] A.Klimishin, *Shock waves in stellar atmospheres* (in Russian), Nauka, Moscow 1984

[83] Ya.B.Zel'dovich, Yu.P.Raizer, *The physics of shock waves and high temperature hydrodynamic phenomena* (in Russian), Nauka, Moscow 1966, Engl. Transl.: Academic Press, New York 1967

[84] C.K.Chu, R.A.Gross, *Shock waves in plasma physics*, in: *Advances in plasma physics*, ed. by A.Simon, W.B.Thompson, vol.2, Interscience Publishers, New York 1969

[85] J.E.Anderson, *Magnetohydrodynamic shock waves*, MIT Press, Cambridge 1963

[86] M.A.Liberman, A.L.Velikovich, *Physics of shock waves in gases and in plasmas*, Springer-Verlag, Berlin, Heidelberg, New York, Tokyo 1986

[87] H.R.Griem, *Validity of local thermodynamic equilibrium in plasma spectroscopy*, Phys. Rev. 131 no.3 (1963) 1170-1176

[88] H.W.Drawin, *Validity conditions for local thermodynamic equilibrium*, Z.Physik 228 (1969) 99-119

[89] H.W.Drawin, *Bedingungen für die Gültigkeit des lokalen thermodynamischen Gleichgewichts*, in: *Ergebnisse der Plasmaphysik und Gaselektronik*, vol. III, Akademie-Verlagm, Berlin 1972, pp.363-434

[90] W.A.Cillers, J.D.Hey, J.P.S.Rash, *Thermal equilibrium criteria in a nitrogen plasma*, J. Quant. Spectr. Radiation Transfer 15 (1975) 963-978

[91] W.Neumann, *Der Katodenmechanismus von Hochdruckbögen*, Beitr. Plasmaphys. 9 no.6 (1969) 499-526

[92] W.Neumann, *The mechanism of the thermo-emitting arc cathode*, Akademie-Verlag, Berlin 1987

[93] Ch.Weißmantel, C.Hamann, *Grundlagen der Festkörperphysik*, Deutscher Verlag der Wissenschaften, Berlin 1979

[94] W. Kluge, *Glühemission und Austrittsarbeiten*, in: Landoldt-Börnstein, *Zahlenwerte und Funktionen aus Naturwissenschaft und Technik*, Band 2, Teil 6, Springer-Verlag Berlin, Göttingen, Heidelberg 1959

[95] S.S.Mackeown, *The cathode drop in an electric arc*, Phys. Rev. vol.34 Ser.2 no.4 (1929) 611-614

[96] R.H.Fowler, L.Nordheim, *Electron emission in intense electric fields*, Proc. Roy. Soc. A119 (1929) 173

[97] J.Dobrogowski, H.Moscicka-Grzesiak, St.Seidel, Z.Załucki, *Fundamentals of discharges and electrical insulation in vacuum* (in Polish), Panstwowe Wydawnictwo Naukowe, Warsaw 1983

[98] R.H. Good Jr., E.W.Müller, *Field emission*, in: Handbuch der Physik vol.XXI, Springer-Verlag, Berlin, Göttingen, Heidelberg, 1956, 176-231

[99] E.L.Murphy, R.H.Good Jr., *Thermionic emission, field emission and the transition region*, Phys. Rev. 102 no.6 (1956) 1464-1473

[100] T.H.Lee, *T-F theory of electron emission in high-current arcs*, J.Appl. Phys. 30 no.2 (1959) 166-171

[101] G.A.Mesyats, D.I.Proskurovskii, *Pulsed electrical discharge in vacuum* (in Russian), Nauka, Novosibirsk 1984; Engl. Transl.: Springer-Verlag Berlin 1989 (Springer Series in Atoms and Plasmas, vol.5)

[102] M.Kaminsky, *Atomic and ionic impact phenomena on metal surfaces*, Springer-Verlag Berlin, Heidelberg, New York 1965

[103] G.V.Marr, *Plasma spectroscopy*, Elsevier Publishing Company, Amsterdam, London, New York 1968

[104] A.Möschwitzer (Editor), *Formeln der Elektrotechnik und Elektronik*, Verlag Technik, Berlin, and Carl Hanser Verlag, München, Wien, 1986

[105] A.B.Gurevich, L.P.Pitaevskii, *Recombination coefficient in a dense, low-temperature plasma*, Zh. Eksp. Teor. Fiz. 46 no.4 (1964) 1281-1284; Sov. Phys. JETP 19 no.4 (1964)

[106] L.P.Pitaevskii, *Recombination of electrons in a monoatomic gas*, Zh. Eksp. Teor. Fiz. 42 no.5 (1962) 1326-1329

[107] A.Einstein, *Über die von der molekular-kinetischen Theorie der Wärme geforderte Bewegung von in ruhenden Flüssigkeiten suspendierten Teilchen*, Annalen Physik (Leipzig) 17 (1905) 549-560

[108] L.G.H.Huxley, R.W.Crompton, M.T.Elford, *Use of the parameter* E/N, Br. J. Appl. Phys. 17 (1966) 1237-1238

[109] A.Fick, *Über Diffusion*, Annalen Phys. Chem. 94 (1955) 59-86

[110] E.A.Mason, E.W.McDaniel, *Transport properties of ions in gases*, Wiley Interscience, New York 1988

[111] G.H.Wannier, *Motion of gaseous ions in strong electric fields*, Bell. Syst. Tech. J. 32 (1953) 170-254

[112] L.G.H.Huxley, R.W.Crompton, *The diffusion and drift of electrons in gases*, Wiley, New York 1974, Russian transl.: Mir, Moscow, 1977

[113] T.Kihara, *The mathematical theory of electrical discharges in gases: B. Velocity distribution of positive ions in a static field*, Rev. Mod. Phys. 25 (1953) 844-852

[114] E.W.McDaniel, *Collision phenomena in ionized gases*, John Wiley & Sons, New York, London, Sydney 1964, Russian transl.: Mir, Moscow, 1967

[115] L.B.Loeb, *Basic processes of gaseous electronics*, University of California Press, Berkeley and Los Angeles 1960

[116] W.P.Allis, *Motion of ions and electrons*, in: Handbuch der Physik XXI pt.I, Springer-Verlag, Berlin 1956, pp. 383-444

[117] W.Schottky, *Diffusionstheorie der positiven Säule*, Phys. Zeitschrift 25 (1924) 635-640

[118] A.A.Radtsig, B.M.Smirnov, *Parameters of atoms and atomic ions* (in Russian), 2nd ed., Energoatomizdat Moscow 1986

[119] N.E.Holden, R.L.Martin, *Atomic weights of the elements*, Pure Appl. Chem. 56 (1984) 653-674

[120] E.U.Condon, H.Odabasi, *Atomic structure*, Cambridge Univ. Press, London 1980

[121] I.I.Sobelman, *Introduction to the theory of atomic spectra* (in Russian), Fizmatgiz, Moscow 1963, 2nd. ed. Nauka Moscow 1977, Engl. transl.: *Atomic spectra and radiative transitions*, Springer-Verlag, Berlin, Heidelberg, New York 1979

[122] Grimsehl, *Lehrbuch der Physik*, Vol.II, 13.ed., Teubner Verlagsgesellschaft, Leipzig, 1954

[123] E.Hantzsche, *The thermo-field emission of electrons in arc discharges*, Beitr. Plasmaphys. 22 no.4 (1982) 325-346

[124] S.G.Christov, *General theory of electron emission from metals*, Phys. Stat. Solidi 17 (1966) 11-26

[125] D.Hinschelwood, *Explosive emission cathode plasma in intense relativistic electron beam diode*, NRL Memorandum Report 5492, Naval Research Laboratory, Washington 1985

[126] I.M.Bronshtein, B.S.Fraiman, *Secondary electron emission* (in Russian), Nauka, Moscow 1969

[127] S.Anders, A.Anders, *On modes of arc cathode operation*, Preprint Zentralinstitut für Elektronenphysik (ZIE), Berlin 1990; to be published in IEEE Trans. Plasma Science 18 no.5 (1990)

[128] K.Miyamoto, *Plasma physics for nuclear fusion*, The MIT Press, Cambride, 1980

[129] D.J.Rose, M.Clark Jr., *Plasma physics and controlled fusion*, The MIT Press and John Wiley & Sons, New York and London

[130] W.M.Stacey Jr., *Fusion. An introduction to the physics and technology of magnetic confinement fusion*, Wiley Interscience, New York 1984

[131] J.Wesson, *Tokamaks*, Clarendon Press, Oxford 1987

[132] P.M.Chung, L.Talbot, K.J.Touryan, *Electric probes in stationary and flowing plasmas: theory and application*, Springer-Verlag, Berlin, Heidelberg, New York, 1975; Russian transl.: Mir, Moscow, 1978

[133] L.Tonks, I.Langmuir, *A general theory of the plasma of an arc*, Phys. Rev. vol.34 Ser.2 no.4 (1929) 876-922

[134] D.Bohm, E.H.S.Burhop, H.S.W.Massey, *Use of probes for plasma exploration*, in: *The characteristics of electrical discharges in magnetic fields*, ed. by A.Guthrie and R.K.Wakerling, McGraw-Hill, New York, chpt.2

[135] F.F.Chen, *Electrical probes*, in: *Plasma diagnostic techniques*, ed. by R.H.Huddelstone, S.L.Leonard, Academic Press, New York, London, 1965, chpt.4; Russian transl.: Mir, Moscow, 1967

[136] S.Ullrich, *Plasmaphysikalische Grundlagen und Methoden des Einsatzes von Plasmasonden in der Raumforschung*, Akademie-Verlag Berlin, 1976

[137] O.V.Kozlov, *Electrical probe in a plasma* (in Russ.), Atomizdat Moscow, 1969

[138] J.D.Swift, M.J.R.Schwar, *Electrical probes for plasma diagnostics*, Iliffe Books Ltd, London, 1970

[139] B.V.Alekseev, V.A.Kotel'nikov, *The probe method in plasma diagnostics* (in Russ.), Energoatomizdat, Moscow 1988

[140] M.S. Benilov, *Theory of electrical probes in flowing weakly ionized high-pressure plasmas* (Review), Teplofiz. Vys. Temperatur 26 no.5 (1988) 993-1004

[141] L.Schott, *Plasma diagnostics using electrical probes*, in: *Reactions under plasma conditions*, ed. by M.Venugopalan, Wiley Interscience, New York, 1971, vol.1, 515-542

[142] C.H.Su, S.H.Lam, *Continuum theory of spherical electrostatic probes*,Phys. Fluids 6 no.10 (1963) 1479-1491

[143] R.E.Kiel, *Continuum electrostatic probe theory for large sheaths on spheres and cylinders*, J.Appl.Phys. 40 no.9 (1969) 3668-3673

[144] P.M.Chung, *Electrical characteristics of Couette and stagnation boundary-layer flows of weakly ionized gases*, Phys. Fluids 7 no.1 (1964) 110-120

[145] A.J.Russo, K.J. Touryan, *Experimental and numerical studies of flush electrostatic probes in hypersonic ionized flows*, AIAA J. 10 no.12 (1972) 1675-1678

[146] G.K. Bienkowski, *Electrostatic sheath in a weakly ionized gas*, Phys. Fluids 10 no.2 (1967) 381-391

[147] R.A.Alpher, D.R.White, *Optical interferometry*, chpt.10 of: *Plasma diagnostic techniques*, ed. by R.H.Huddlestone and S.L.Leonard, Academic Press, New York, 1965; Russian Transl.: Mir, Moscow, 1967

[148] W.L.Wiese, *Breadth of spectral lines*, ibid chapter 6

[149] R.W.P.McWhriter, *Spectral intensities*, ibid chapter 5

[150] G.V.Ostrovskaya, Yu.I.Ostrovskii, *Holographic methods of plasma diagnostics*, Progress in Optics (ed. by E.Wolf) 22 (1985) 199-270

[151] Yu.I.Ostrovskii, *Holography and its application* (in Russ.), Nauka, Leningrad, 1972, 1976, 1982, 1986; German Transl.: Teubner Verlagsgesellschaft, Leipzig, 1987

[152] M.F.Kimmitt, A.C.Prior, V.Roberts, *Measurments in the infrared region*, as Ref. [147]: chapter 9

[153] T.F.Stratton, *X-ray spectroscopy*, as Ref. [147]: chapter 8

[154] S.L.Leonard, *Basic macroscopic measurments*, as Ref. [147]: chapter 1

[155] A.Ovsynnicov, *Microwave and laser diagnostics of plasmas*, chapter 8 of: *Reactions under plasma conditions*, vol.I, ed. by M.Venugopalan, Wiley-Interscience, New York, London, Sydney, Toronto, 1971

[156] A.Giraud, *Ion mass-spectrometry*, chapter 10 of: *Reactions under plasma conditions*, vol.I, ed. by M.Venugopalan, Wiley-Interscience, New York, London, Sydney, Toronto, 1971

[157] S.Glasstone, R.H.Lovberg, *Controlled thermonuclear reactions*, Van Nostrand, New York, 1960

[158] I.P.Grant, *Calculation of Gaunt factors for free-free transitions near positive ions*, Mont. Not. Roy. Astron. Soc. 118 no.3 (1958) 241-257

[159] W.J.Karzas, R.Latter, Astrophys. J. Suppl. 6 (1961) 167

[160] D.Schlüter, *Gaunt factors for free-free transitions in the field of positive ions of carbon, nitrogen, and oxygen, calculated in the Scaled Thomas-Fermi approximation*, Z. Phys. D - Atoms, Molecules and Clusters 11 (1989) 273-276

[161] *Bibliography on atomic transition probabilities*, NBS Special Publication 505, Washington, 1978; NBS Special Publication 505, Supplement 1, Washington 1980

[162] compiled by D.Sünder, Central Institute of Electron Physics, Berlin

[163] S.I.Braginskii, *Plasma transport phenomena* (in Russian), Voprosy Teorii Plazmy, Gosatomizdat, Moscow 1963, pp.183-272; Engl. Transl.: *Transport processes in a plasma*, Review of Plasma Physics, vol.1, Consultants Bureau, New York 1965

[164] A.Garscadden, *Gas discharge laser diagnostics update*, in: *Radiative Processes in discharge plasmas*, ed. by J.M. Proud and L.H.Luessen, NATO ASI Series, Series B: Physics, vol.149, pp.467-494, Plenum Press, New York, 1986

[165] D.R.Crosley, G.P.Smith, *Laser-induced fluorescence spectroscopy for combustion diagnostics*, Optical Engineering 22 no.5 (1983) 545-553

[166] C.A.Moore, G.P.Davis, R.A.Gottscho, *Sensitive, nonintrusive, in-situ measurments of temporeally and spatially resolved plasma electric field*, Phys. Rev. Lett. 52 no.7 (1984) 538-541

[167] P. van der Weijer, R.M.M.Cremers, *Experimental determination of* $6\ ^3P - 6\ ^3P$ *collisional excitation cross section for line emission in the positive column of dc mercury discharge*, J. Appl. Phys. 53 no.3 (1982) 1401-1408

[168] C.E.Gaebe, R.A.Gottscho, *Laser (and other) diagnostics of* rf *discharges*, in: *Radiative Processes in discharge plasmas*, ed. by J.M. Proud and L.H.Luessen, NATO ASI Series, Series B: Physics, vol.149, pp.495-523, Plenum Press, New York, 1986

[169] J.E.Lawler, D.K.Doughty, E.A. Den Hartog, S.Salih, *Optogalvanic effects in the cathode fall*, in: *Radiative Processes in discharge plasmas*, ed. by J.M. Proud and L.H.Luessen, NATO ASI Series, Series B: Physics, vol.149, pp.525-546, Plenum Press, New York, 1986

[170] compiled under cooperation with T. Kahlbaum, Central Institute of Electron Physics, Berlin; see also: T.Kahlbaum, A.Forster, *Thermodynamic properties of nonideal plasmas with multiple ionization and Coulomb & hard-core interactions*, Laser Particle Beams (1990) in press

[171] P.Debye, E.Hückel, Phys. Zeitschrift 24 (1923) 185

[172] A.Unsöld, *Zur Berechnung der Zustandsummen für Atome und Ionen in einem teilweise ionisiertem Gas*, Zeitschrift Astrophys. 24 (1949) 355-362

[173] G.Ecker, W.Kröll, *Lowering of the ionization energy for a plasma in thermodynamic equilibrium*, Physics of Fluids 6 no.1 (1963) 62-70

[174] J.A.Gaunt, *Continuous absorption*, Phil. Trans. Roy. Soc. (London) A229 (1930) 163

[175] R.Kollath, *Sekundärelektronenemission fester Körper bei Bestrahlung mit Electronen*, in: Handbuch der Physik, vol.21, Springer-Verlag Berlin, 1956, 282-303;
R.Kollath, *Sekundärelektronenemission fester Körper*, in: Landoldt-Bornstein, *Zahlenwerte und Funktionen aus Naturwissenschaft und Technik*, vol.2, Springer-Verlag, Berlin, Göttingen, Heidelberg, 1959, 1004-1018

[176] H.L.Anderson (Editor in Chief), *A physicist's desk reference* (The second edition of *Physics vade mecum*), American Instiute of Physics, New York, 1989

[177] K.R.Lang, *Astrophysical formulae (A compendium for the physicist and astrophysicist)*, 2nd corrected and enlarged edition, Springer-Verlag, Berlin, Heidelberg, New York, 1980

[178] W.Lindinger, T.D.Märk, F.Howorka (Editors), *Swarms of ions and electrons in gases*, Springer-Verlag, Wien, New york, 1984

[179] J.Hinze (Editor), *Electron-atom and electron-molecule collisions*, Plenum Press, New York, London, 1983

[180] B.D.Fried, S.C.Conte, *The plasma dispersion function*, Academic Press, New York, 1961

[181] R.W.Landau, S.Cuperman, *Stability of anisotropic plasmas to almost-perpendicular magnetosonic waves*, J. Plasma Phys. 6 pt.3 (1971) 495-512

[182] B.D.Fried, C.L.Hedrick, J.McCune, *Two-pole approximation for the plasma dispersion function*, Phys. Fluids 11 no.1 (1968) 249-252

[183] H.Paul, *Lasertheorie*, 2 volumes, Akademie-Verlag Berlin, Pergamon Press Oxford, Vieweg Braunschweig, 1969

[184] D.Röss, *Laser light amplifieres and oscillators*, Academic Press, London and New York, 1969

[185] C.Willett, *Introduction to gas lasers, population inversion mechanism with emphasis on selective excitation processes*, Pergamon Press, Oxford, 1974

[186] W.W.Duley, *Laser processing and analysis of materials*, Plenum Press, New York and London, 1983; Russian transl.: Mir Moscow 1986

[187] W.Brunner, K.Junge et al., *Wissensspeicher Lasertechnik*, Fachbuchverlag, Leipzig, 1982

[188] Ch.K.Rhodes (Editor), *Excimer lasers*, Second Enlarged Edition, Springer-Verlag, Berlin, Heidelberg, New York, Tokyo, 1984

[189] C.D.Child, Phys. Rev. 32 (1911) 492-511; I.Langmuir, Phys. Zeitschrift 15 (1914) 348-353

[190] H.-P. Stormberg, *Line broadening and radiative transport in high- pressure mercury discharges with* Na I *and* Tl I *as additives*, J. Appl. Phys. 51 no. 4 (1980) 1963-1969

[191] D.Mihalas, B.Weibel-Mihalas, *Foundations of radiation hydrodynamics*, Oxford University Press, New York and London, 1984

[192] G.B.Rybicki, A.P.Lightman, *Radiative processes in astrophysics*, John Wiley & Sons, New York, 1979

[193] Yu.S.Protasov, *Thermodynamic and optical properties of ionized gases for temperatures up to* 100 *eV* (in Russian), Atomizdat, Moscow, 1988

[194] B.N.Tsetverushkin, *Mathematical modelling of problems in radiation gas dynamics* (in Russian), Nauka, Moscow, 1985

[195] I.I.Sobelman, L.A.Vainstein, E.A.Yukov, *Excitation of atoms and broadening of spectral lines*, Russian original: Nauka, Moscow, 1979, English transl.: Springer-Verlag, Berlin, Heidelberg, New York, 1980

[196] G.J.Pert, *Models of collisional-radiative recombination*, J. Phys. B: At. Mol. Opt. Phys. 23 (1990) 619-650

[197] P.C.Stangeby, *Analysis of bolometer / Langmuir and ion-trapping / Langmuir probe operation*, J. Nucl. Mater. 111 & 112 (1982) 84-86

[198] P.C.Stangeby, *The plasma sheath*, in: *Physics of plasma-wall interactions in controlled fusion*, ed. by D.E.Post and R.Behrisch, NATO ASI Series B: Physics vol. 131, 1985, pp. 41-97

SUBJECT INDEX

A

B

C

D

E

F

G

H

I

J

K

L

M

N

O

P

Q

R

S

T

U

V

W

X

A Formulary for Plasma Physics
presents a collection of Plasma Physics formulae compiled in such a way as to facilitate quick and easy location of the required expression. Dealing not only with classical plasma problems such as plasma waves, radiation and composition but also with problems of "peripheral" importance such as the electronic structure of atoms and electrode phenomena, the formulary has been designed for use by scientists, engineers and students working in Plasma Physics and related fields.
Where appropriate, formulae are accompanied by abbreviated or approximate forms suitable for quick calculation, together with a list of references for those seeking a derivation of the expression. In the majority of cases, formulae are written in both systems of units, *SI* and *CGS*, and an extensive compilation of constants, units and definitions is also included.

ISBN 3-05-501263-1

Zeitfracht Medien GmbH
Ferdinand-Jühlke-Straße 7
99095 Erfurt, Deutschland
produktsicherheit@kolibri360.de